T0140436

Lecture Notes in Networks and Systems

Volume 964

The series "Lecture Notes in Networks and Systems" publishes the latest developments in Networks and Systems—quickly, informally and with high quality. Original research reported in proceedings and post-proceedings represents the core of LNNS.

Volumes published in LNNS embrace all aspects and subfields of, as well as new challenges in, Networks and Systems.

The series contains proceedings and edited volumes in systems and networks, spanning the areas of Cyber-Physical Systems, Autonomous Systems, Sensor Networks, Control Systems, Energy Systems, Automotive Systems, Biological Systems, Vehicular Networking and Connected Vehicles, Aerospace Systems, Automation, Manufacturing, Smart Grids, Nonlinear Systems, Power Systems, Robotics, Social Systems, Economic Systems and other. Of particular value to both the contributors and the readership are the short publication timeframe and the world-wide distribution and exposure which enable both a wide and rapid dissemination of research output.

The series covers the theory, applications, and perspectives on the state of the art and future developments relevant to systems and networks, decision making, control, complex processes and related areas, as embedded in the fields of interdisciplinary and applied sciences, engineering, computer science, physics, economics, social, and life sciences, as well as the paradigms and methodologies behind them.

Indexed by SCOPUS, INSPEC, WTI Frankfurt eG, zbMATH, SCImago.

All books published in the series are submitted for consideration in Web of Science.

For proposals from Asia please contact Aninda Bose (aninda.bose@springer.com).

Debasis Giri · Jaideep Vaidya · S. Ponnusamy ·
Zhiqiang Lin · Karuna Pande Joshi ·
V. Yegnanarayanan
Editors

Proceedings of the Tenth International Conference on Mathematics and Computing

ICMC 2024, Volume 1

Springer

Editors
Debasis Giri
Department of Information Technology
Maulana Abul Kalam Azad University
of Technology
Kolkata, West Bengal, India

S. Ponnusamy
IIT Madras
Chennai, Tamil Nadu, India

Karuna Pande Joshi
The University of Maryland, Baltimore
County (UMBC)
Baltimore, USA

Jaideep Vaidya
Rutgers University
New Jersey, USA

Zhiqiang Lin
The Ohio State University
Columbus, USA

V. Yegnanarayanan
Kalasalingam Academy of Research
and Education
Srivilliputhur, Tamil Nadu, India

ISSN 2367-3370   ISSN 2367-3389 (electronic)
Lecture Notes in Networks and Systems
ISBN 978-981-97-2065-1   ISBN 978-981-97-2066-8 (eBook)
https://doi.org/10.1007/978-981-97-2066-8

Committees

Chief Patron

Dr. K. Sridharan, Chancellor, Kalasalingam Academy of Research and Education, Tamil Nadu, India

Patrons

Smt. Dr. S. Arivalagi, Pro-Chancellor, Kalasalingam Academy of Research and Education, Tamil Nadu, India
Dr. S. Shasi Anand, Vice-President, (Academic), Kalasalingam Academy of Research and Education, Tamil Nadu, India
Er. S. Arjun Kalasalingam, Vice-President (Administration), Kalasalingam Academy of Research and Education, Tamil Nadu, India

Co-patrons

S. Narayanan, Vice Chancellor, Kalasalingam Academy of Research and Education, Tamil Nadu, India
V. Vasudevan, Registrar, Kalasalingam Academy of Research and Education, Tamil Nadu, India

General Chairs

P. K. Saxena, Scientific Consultant-Cyber Security, Government of India, Former Director, SAG, DRDO, New Delhi, India
P. D. Srivastava, IIT Bhilai, Raipur, India

Programme Chairs

Jaideep Vaidya, Rutgers University, USA
S. Ponnusamy, IIT Madras, Chennai, India
Zhiqiang Lin, The Ohio State University, USA
Karuna Pandey Joshi, The University of Maryland, Baltimore County (UMBC), USA
Debasis Giri, Maulana Abul Kalam Azad University of Technology, WB, India

Organizing Chair

Yegnanarayanan Venkataraman, Kalasalingam Academy of Research and Education, TN, India

International Advisory Committee

Alfred Menezes, University of Waterloo, Canada
Bhavani Thuraisingham, University of Texas, USA
Bimal Roy, ISI Kolkata, India
Chris Mitchell, Royal Holloway, University of London, UK
Dieter Gollmann, Hamburg University of Technology, Germany
Elisa Bertino, Purdue University, USA
Heinrich Begehr, Freie Universitat Berlin, Germany
Joel J. P. C. Rodrigues, University of Petroleum (East China), China
Kouichi Sakurai, Kyushu University, Japan
Lajos Hanzo, University of Southampton, Southampton, UK
Mahesh Kalyanaraman, Associate Vice-President, HCL, India
Mark Sepnaski, Baylor University, Waco, Texas, USA
Mohan Kankanhalli, National University of Singapore, Singapore
Mohammad S. Obaidat, University of Texas-Permian Basin, USA
Merouane Debbah, University of Texas-Permian Basin, USA
Moti Yung, Columbia University, USA
Oscar Castillo, Tijuana Institute of Technology, Mexico

Rajkumar Buyya, University of Melbourne, Australia
Rakesh M. Verma, University of Houston, USA
Sokratis Katsikas, Norwegian University of Science and Technology-NTNU, Norway
Valentina E. Balas, Aurel Vlaicu University of Arad, Romania
Veni Madhavan C. E., IISc Bangalore, India
Vishnu S. Pendyala, San Jose State University, USA

Members of the Organizing Committee

P. Deepalakshmi, Dean, SOC, KARE
C. Ramalingan, Dean, SAS, KARE
P. Sarasu, Director—International Relations, KARE
K. Karuppasamy, Professor of Mathematics, KARE
S. Balamurali, Senior Professor of Mathematics, KARE
S. Dhanasekaran, HOD, IT, KARE
K. Kartheeban, Head, CA and CSIT, KARE
M. Kameswari, HOD, Maths, KARE
K. Suthendran, Associate Professor (SG), IT, KARE
M. Jayalakshmi, Associate Professor, CSE, KARE
L. Sreenivasulu Reddy, Associate Professor, Maths, KARE
B. Reddappa, Associate Professor, Maths, KARE
T. Shanmughapriya, Assistant Professor, Maths, KARE
N. Sureshkumar, Professor, CSE, KARE
C. Balasubramanian, Associate Professor, CSE, KARE
N. C. Brintha, Associate Professor, CSE, KARE
Amutha Guka, Associate Professor, CA, KARE
V. Baby Shalini, Associate Professor, IT, KARE
K. Maharajan, Associate Professor, CSE, KARE
Jane Rubel Angelina, Associate Professor, CSE, KARE
R. Murugeswari, Associate Professor, CSE, KARE
T. Sam PradeepRaj, Associate Professor, CSE, KARE
A. Parivazhagan, Associate Professor, CSE, KARE
V. Anusuya, Associate Professor, CSE, KARE
Abhishek Tripathi, Associate Professor, CSE, KARE

Members of the Design and Web Development

S. Suprakash and Prem Raja, School of Computing and his Team

Members of the Technical Programme Committee

Abhishek Kumar Singh, VIT University Chennai, India
Achyutha Krishnamoorthy, CMS College Kottayam, Kerala, India
A. Menaka Pushpa, VIT, Chennai, India
A. Vanav Kumar, National Institute of Technology, Arunachal Pradesh, India
Aditi Gangopadhyay, IIT Roorkee, India
Amit Prakash Singh, Guru Gobind Singh Indraprastha University, New Delhi, India
Amitabh Gyan Ranjan, MGCUB, Mahatma Gandhi Central University, Bihar, India
Amrit Pal, VIT, Chennai, India
Anita Pal, National Institute of Technology Durgapur, India
Anjana Gosain, Guru Gobind Singh Indraprastha University, Punjab, India
Anup Kumar Sharma, NIT Raipur, India
Anurag Goel, Delhi Technological University, New Delhi, India
Apu Kumar Saha, NIT Agartala, Agartala, Tripura, India
C. Arun Kumar, Amrita Vishwa Vidyapeetham, Coimbatore, India
Arvind K. R. Sinha, NIR, Raipur, India
Arya Kumar Bedabrata Chand, IIT Madras, India
Ashok Kumar Das, University of Calcutta, West Bengal, India
R. V. Aswiga, VIT, Chennai, India
Ajanta Das, Amity University Kolkata, India
Ali Ebrahimnejad, Qaemshahr Branch, Islamiz Azad University, Qaemshahr, Iran
Arif Ahmed Sk, University of Tromsø, Norway
Arvind Kumar Gupta, IIT Ropar, Punjab, India
Amitabh Gyan Ranjan, MGCUB, Mahatma Gandhi Central University, Bihar, India
Arshad Khan, Jamia Millia Islamia (A Central University) New Delhi, India
Abhijit Das, NIT Trichy, Thiruchirapalli, India
Ankita Vaish, Institute of Science Banaras Hindu University, India
Arvind Selwal, Central University of Jammu, India
Aquil Khan, Indian Institute of Technology Indore, India
A. Swaminathan, IIT Roorkee, India
Arvind Selwal, Central University of Punjab Bathinda, India
Atul, NIT, Trichy, India
Aditi Gangopadhyay, IIT Roorkee, India
A. K. Bedabrata Chand, IIT Madras, India
Amandeep Kaur, Central University of Jammu, India
Anshul Verma, Institute of Science, Banaras Hindu University, Varanasi, India
Abhijit Datta Banik, IIT Bhubaneswar, India
Ajit Das, Bodoland University Kokrajhar, Assam, India
Ajoy Kumar Khan, Mizoram University Aizawl, Mizoram, India
Aleksandr Poliakov, Sevastopol State University, India
Amit Banerjee, South Asian University, New Delhi, India
Amit Maji, IIT Kharagpur, India
Amiya Nayak, University of Ottawa, Canada

Ankit Rajpal, University of Delhi, Delhi, India
Anuj Jakhar, IIT Madras, India
Anupam Saikia, Indian Institute of Technology Guwahati, India
Arnab Patra, Indian Institute of Technology Kharagpur, India
Arun Kumar, National Institute of Technology, Rourkela, Odisha, India
A. Swaminathan, IIT Roorkee, India
Asish Bera, BITS Pilani, India
Atena Ghasemabadi, Esfarayen University of Technology, India
Ayaz Ahmad, NIT Patna, India
Ashok Kumar Das, International Institute of Information Technology, Hyderabad, India
Anoop Singh, IIT (BHU), Varanasi, India
Atasi Deb Ray, University of Calcutta, India
S. Ananda Kumar, VIT, Chennai, India
Annappa, NIT Karnataka, Surathkal, India
A. Senthil Thilak, NIT Karnataka, Surathkal, India
Alagu Manikandan Esakkimuthu, Cognizant, Hartford, USA
Adhvitha Premanand, NUS-National University of Singapore, Singapore
Anantha Narayanan, Accenture Technology Solutions, Dallas, USA
Arunkumar Pichaimuthu, Verizon, Concord, USA
Andavarayan Muthuraju, Syntel Ltd., Toronto, Canada
Anabayan, Cap Gemini, Atlanta, USA
Arun Kumar, Langara College, Vancouver, BC Richmond, Canada
Anandaraj Selvarangan, Tata Consultancy Services, Minneapolis, USA
Amutha, ABB Ltd., California City, USA
Akhil, Ericsson, San Francisco, USA
J. Abraham Stephen, TCS, Plano, USA
Arthy Paul Brito, Cardiff University, Wales, UK
Amit Kumar Verma, IIT Patna, India
Avijit Duary, Maulana Abul Kalam Azad University of Technology, West Bengal, India
Adriana Mihaela Coroiu, Babes-Bolyai University, Romania
Abd Raouf ChouikhaView Abd, Universite Paris-Sorbonne, Paris Nord, France
Angela Thompson, Governors State University, Greater Chicago, USA
Andrew Pownuk, The University of Texas at El Paso, USA
Abhimanyu Mukerji, Amazon, Bay Area, California, USA
Ashwin Viswanathan Kannan, Amazon Labs CA, USA
Balachander Palanisamy, Computer Sciences Corporation, USA
Balaji Krishnamoorthy, NTT Data UK, Bristol, UK
Bimal Mandal, IIT Jodhpur, India
Bipul Kumar Sarmah, Tezpur University, Napaam, Sonitpur, Assam, India
Bok-Min Goi, Universiti Tunku Abdul Rahman, Malaysia
Baskar Babujee, MIT Campus, Anna University, India
B. Sundaravadivoo, Alagappa University, Karaikudi, India
Bapan Ghosh, Indian Institute of Technology Indore, India

B. Nithya, National Institute of Technology Tiruchirappalli, India
Balagopal Komarath, IIT Gandhinagar, Gujarat, India
Balendu Bhooshan Upadhyay, IIT Patna, Bihar, India
Bankim C. Mandal, IIT Bhubaneswar, India
Bhabesh Nath, Tezpur University, India
Bhanuman Barman, National Institute of Technology Patna, Bihar, India
Bidyut Kr. Patra, National Institute of Technology, India
Bimal Roy, Indian Statistical Institute, Kolkata, India
Bipul Sauraph, IIT Gandhinagar, Gujarat, India
B. R. Shankar, NITK (National Institute of Technology Karnataka), Surathkal, Mangalore, India
B. Sundaravadivoo, Alagappa University, Karaikudi, India
Balasubramanian Raman, Indian Institute of Technology Roorkee, India
Balendu Bhooshan Upadhyay, IIT Patna, Bihar, India
Bijendra Kumar, Nethaji Subhas University of Technology, New Delhi, India
C. Mandal, IIT Kharagpur, West Bengal, India
C. Chandra Sekhar, IIT Madras, India
J. Cynthia, Kumaraguru College of Technology, Coimbatore, India
Costin Badica, University of Craiova, Romania
Ch. Srinivasa Rao, IIT Madras, India
C. Shoba Bindu, JNTU Anantapur, India
A. Chandrashekaran, Central University of Tamil Nadu, India
Chayan Halder, West Bengal State University, India
Chien-Ming Chen, Harbin Institute of Technology Shenzhen Graduate School, China
Carynthia Kharkongor, Delhi Skill and Entrepreneur University, France
K. Chandrasekaran, NIT Karnataka, Suratkal, France
Christina Boura, Université de Versailles Saint-Quentin-en-Yvelines, France
Chandramohan Lakshman, TriNet, PeopleSoft Technical Professional HCM/CRM, San Francisco, USA
Chandrashekhar Ramamurthi, Verizon Telecom, Dallas, USA
Canan Bozkaya, Middle East Technical University, Turkey
Diwas Srivastava, CNRS, Montpellier, France
Dhamodharan Narayanan, AT&T, Telecommunications, Dallas, USA
R. Divya Meenakshi, Qode Media Inc., Canada
Daya Gupta, Delhi Technological University, New Delhi, India
Digambar Pawar, University of Hyderabad, France
D. Mishra, NIT Raipur, India
Dalibor Froncek, University of Minnesota, USA
Debasis Giri, Maulana Abul Kalam Azad University of Technology, West Bengal, India
Deepa Kumar, Kalasalingam Academy of Research and Education, Tamil Nadu, India
Deepjyoti Goswami, Tezpur University, Napaam, Assam, India
Dieter Gollmann, Hamburg University of Technology, Germany
Dijana Mosic, University of Nis, Serbia

Dilip Kumar Maiti, Vidyasagar University, West Bengal, India
Dilpreet Kaur, Indian Institute of Technology Jodhpur, India
Djamal Foukrach, Hassiba Benbouali University of Chlef, Algeria
D. Bhargavi, NIT Warangal, India
Debashis Dutta, NIT Warangal, India
Debashis Nandi, National Institute of Technology, India
Debraj Giri, IIT, India
Deepak Ranjan Nayak, Malaviya National Institute of Technology Jaipur, Rajasthan, India
P. Deepalakshmi, Kalasalingam Academy of Research and Education, India
Devanayagam Palaniappan, Texas A&M University Corpus—Christi, USA
Dhananjoy Dey, Indian Institute of Information Technology Lucknow, India
Dinabandhu Pradhan, IIT (ISM) Dhanbad, India
Dipanwita Roy Chowdhury, IIT Kharagpur, India
Durga Charan Dalal, Indian Institute of Technology Guwahati, India
Debasish Bera, Indian Institute for Information Technology, Kalyani, India
Debasisha Mishra, NIT Raipur, India
Deepti Jain, Sri Venkateswara College, University of Delhi, India
C. Devadas Nayak, M.I.T., Manipal, India
Dhananjay Gopal, Guru Ghasidas Vishwavidyalaya (A Central University), Bilaspur (C.G.), India
Dharmendra Tripathi, National Institute of Technology, Uttarakhand, India
Edythe E. Weeks, Esq, Washington University, St. Louis, MO, USA
Floyd B. Hanson, University
    of Illinois, USA
Fateme Shahsavand, Malayer University, Malaysia
Fabienne Chaplais, Mathedu SAS, Paris, France
Fagen Li, University of Electronic Science and Technology of China, China
Fu-Hsing Wang, Chinese Culture University, China
Falguni Roy, NIT, Karnataka, Suratkal, India
A. Firos, Rajiv Gandhi University, Arunachal Pradesh, India
Gaurav Bhatnagar, Indian Institute of Technology Jodhpur, China
K. A. Germina, Central University of Kerala, Kerala, India
Gopal Chandra Shit, Jadavpur University, Kolkata, India
Gautham Singh, NIT, Trichy, India
Gururaj Mukarambi, Central University of Karnataka, India
Guru Prem Prasad, IIT Guwahati, India
G. S. Mahapatra, NIT, Puducherry, India
G. Malathi, VIT Chennai, India
Geetanjali Chattopadhyay, Malaviya National Institute of Technology Jaipur, India
T. Gireesh Kumar, Amrita Vishwa Vidyapeetham, Coimbatore, India
Gitanjali Chandwani, IIT Kharagpur, India
Gopal Chandra Shit, Jadavpur University, Kolkata, India
S. Gowrisankar, NIT, Patna, India
H. J. Gowtham, Manipal Academy of Higher Education, Manipal, India

Ganapathy Shanmugam Paramasivan, Medical University of Graz, Graz, Austria
Gaurav Kumar Singh, BE.services GmbH, Munich, Germany
Govindh Sankaran, Wolters Kluwer, Boston, USA
Ganesh Kumar, Judge Software, Charlotte, USA
Gerard Weiss, University of Maastricht, Germany
Halina Kwasnicka, Wroclaw University of Technology, USA
Hisao Ishibuchi, Osaka Prefecture University, Japan
H. Ramesh, Indian Institute of Technology Guwahati, Guwahati, India
Hari Vansh Rai Mittal, Indian Institute of Technology Palakkad, Kerala, India
H. Ramesh, Indian Institute of Technology Guwahati, Guwahati, India
Haowen Tan, Kyushu University, Japan
Hari Vansh RaiMittal, Indian Institute of Technology Palakkad, Kerala, India
Harshdeep Singh, DRDO, New Delhi, India
Heinrich Begehr, Free University Berlin, Germany, Germany
Hiranmoy Mondal, MAKAUT, India
Indranath Sengupta, IIT Gandhinagar, India
Ishapathik Das, IIT Tirupati, AP, India
Indivar Gupta, SAG, DRDO, India
Irani Hazarika, Gauhati University, Guwahati, India
Idrees Qasim, NIT, Srinagar, India
Ibrahim Venkat, Universiti Teknologi Brunei, Malaysia
Ingo Schiermeyer, TU Bergakademie Freiberg, Germany
Jos. M. Koomen, Memes Ltd., Utrecht, The Netherlands
Ji Wang, Ningbo University, China
Jey Veerasamy, The University of Texas at Dallas, USA
Joshva Rajkumar, Ministry for Primary Industries (MPI), Wellington, New Zealand
Joshua Gnanaruban, NCS Pte Ltd. (Singapore), Singapore
Jadav Das, MAKAUT, West Bengal, India
Jamuna Kanta Sing, Jadavpur University, West Bengal, India
Jaspal Singh Aujla, National Institute of Technology, Jalandhar, Punjab, India
JervinZen Lobo, St. Xavier's College Mapusa—Goa, India
Jothi Ramalingam MACS, NIT, Karnataka, Suratkal, India
Jyotsna Yadav, Guru Gobind Singh Indraprastha University, Punjab, India
Jagdish Prasad Maurya, Rajiv Gandhi University (A Central University), Arunachal Pradesh, India
Janakiraman, NIT, Trichy, India
Jaroslaw Adam Miszczak, Institute of Theoretical and Applied Informatics, Polish Academy of Sciences, Poland
Jaydeb Bhaumik, Jadavpur University, Kolkata, India
Jianting Ning, Singapore Management University, Singapore
Jugal K. Verma, IIT Bombay, India
J. Pavan Kumar, National Institute of Technology Tiruchirappalli, India
Jagdish Prakash, University of Johannesburg, South Africa
Jaideep Vaidya, Rutgers University, USA
Jayanta, Tezpur University, Assam, India

Jayanta Mukhopadhyay, Indian Institute of Technology Kharagpur, India
Jothi Ramalingam, MACS, NIT, Karnataka, Suratkal, India
Jugal Prajapat, Central University of Rajasthan, India
Jana Dittmann, Uni Magdeburg, India
Jyotismita Talukdar, Tezpur University, India
J. Christy Roja, St. Joseph's College, Trichy, India
K. Muthukumaran, VIT, Chennai, India
K. Somasundaram, Amrita Vishwa Vidyapeetham, Coimbatore, Tamil Nadu, India
Kartick Chandra Mondal, Jadavpur University, Kolkata, West Bengal, India
Karuna P. Joshi, University of Maryland, Baltimore County, USA
Kaushik Mondal, IIT, Ropar, Punjab, India
Kaushik Roy, West Bengal State University, Barasat, India
Kirankumar R. Hiremath, Indian Institute of Technology Jodhpur, India
Kunwer Singh Mathur, Dr. Hari Singh Gaur Viswavidyalaya Sagar,
Central University of MP, India
Khurram Mustafa, JMI, New Delhi, India
K. Saraswathi, UCEK, JNTUK, India
Kadambari Raghuram, University College of Engineering Kakinada (Autonomous),
Jawaharlal Nehru Technological University Kakinada, India
Kalpesh Kapoor, Indian Institute of Technology Guwahati, India
Kalyanbrat Medhi, Gauhati University, Assam, India
Khaleel Ahmed, Maulana Azad National Urdu University, Telangana, India
Khalid Mahmood, COMSATS University Islamabad, Sahiwal Campus, India
Kotaiah Bonthu, Central Tribal University of Andhra Pradesh, India
Kunwer Singh Mathur, Dr. Hari Singh Gaur Viswavidyalaya Sagar,
Central University of MP, India
K. Palpandi, Malaviya National Institute of Technology (MNIT), Jaipur, Rajasthan,
India
K. Saraswathi, UCEK, JNTUK, India
Kadambari Raghuram, University College of Engineering Kakinada (Autonomous),
Jawaharlal Nehru Technological University Kakinada, India
Kamalika Bhattacharjee, NIT Trichy, India
Khalid Raza, Jamia Millia Islamia (Central University), New Delhi, India
Kolin Paul, IIT Delhi, India
Kushal Sharma, MNIT Jaipur, India
Kusum Sharma, National Institute of Technology, Uttarakhand, India
Karthikeyan Shenbagam, Link Systems, Houston, USA
Kamesh, Infosys, United States
Karthi Vicky, Hinduja Global Solutions, UK
Kouichi Sakurai, Kyushu University, Japan
Kumari Priyanka, University of Hohenheim, Germany
Kavitha Haldorai, Florida State University, USA
K. Somasundaram, Amrita Vishwa Vidyapeetham, Tamil Nadu, India
Longxiu Huang, Michigan State University, India

Lok Pati Tripathi, Indian Institute of Technology Goa Farmagudi, Ponda 403401, Goa, India
Lavanya Selvaganesh, Indian Institute of Technology (BHU), Varanasi, India
Lev Kazakovtsev, Siberian State Aerospace University, Russia
Maharage Nisansala Sevwandi Perera, ATR, Japan
Malay Banerjee, IIT Kanpur, India
Manikandan Rangaswamy, Central University of Kerala, Kerala, India
Manish Kumar Gupta, Guru Ghasidas Vishwavidyalaya, Bilaspur, India
Mario Larangeira, Tokyo Institute of Technology/IOHK, Japan
Md. Abu Talhamainuddin Ansary, IIT-Jodhpur, India
Meenakshi Thakur, Central University of Himachal Pradesh, India
Mohua Banerjee, IIT, Kanpur, India
Moumita Mandal, Indian Institute of Technology Jodhpur, Rajasthan, India
Mriganka Mandal, Indian Institute of Technology (IIT) Jodhpur, India
Muslim Malik, Indian Institute of Technology Mandi, India
Manoj Kumar Singh, Institute of Science, Banaras Hindu University Varanasi, India
Mritunjay Kumar Singh, IIT (ISM) Dhanbad, Dhanbad-826004, Jharkhand, India
Mohd. Arshad, Indian Institute of Technology Indore, Khandwa Road, Simrol, Indore, India
M. P. Pradhan, Sikkim University, Sikkim, India
M. Tiken Singh, Dibrugarh University Institute of Engineering and Technology, India
Madhumangal Pal, Vidyasagar University, India
Mahendra Kumar Gupta, IIT Bhubaneswar, India
Mahipal Jadeja, Malaviya National Institute of Technology, Jaipur, India
Manideepa Saha, NIT Meghalaya, India
Manmohan Vashisth, Indian Institute of Technology (IIT-JMU) Jammu, India
V. Mary Anita Rajam, Anna University, Chennai, India
Md. Obaidullah Sk, Aliah University, India
Megha Khandelwal, Central University of Karnataka, India
Muhammad Abulaish, South Asian University, New Delhi, India
T. Muthukumar, Indian Institute of Technology-Kanpur, India
M. Sethumadhavan, Amrita Vishwa Vidyapeetham, Coimbatore, India
Madhusudana Rao Nalluri, Amrita Vishwa Vidyapeetham, Coimbatore, India
Mahendra Kumar Gupta, IIT Bhubaneswar, India
Mahesh Shirole, VJTI-Mumbai, India
Malay Kule, IIEST, Shibpur, India
Manju Khari, Jawaharlal Nehru University, New Delhi, India
Md. Maqbul, National Institute of Technology Silchar, Assam, India
Mohammad Aslam Siddeeque, AMU, Aligarh, India
Mohammad Mueenul Hasnain, Kamala Nehru College University of Delhi, India
Munesh Meena, DSEU, New Delhi, India
M. Devakar, Visvesvaraya National Institute of Technology (VNIT) Nagpur, India
K. Manikandan, VIT, Vellore, India
Mahendra Pratap Singh, NIT Karnataka, Suratkal, India

Mohammed Hakim Jaffer Ali, SciLifeLab (Science for Life Laboratory), Stockholm, Sweden

Muthukrihnan Govindaraj, eGrove Systems, Sayreville, USA

Manivannan Karunanithi, Endera Systems LLC, McLean, USA

Mariappan Madasamy, HM Revenue and Customs, London, UK

Muthusamy Subash Rajasekaran, Tata Consultancy Services, Bloomington, USA

Mayan Sinha, Salesforce.com, USA

Muthu Rajathi, Tata Consultancy Services Ltd., Manchester, USA

Mohandoss Karuppiah, Cognizant Technology Solutions Pvt. Ltd., Virginia Beach, USA

Mihai Caragiu, Ohio Northern University, USA

Manimuthu Arunmozhi, Aston University, Birmingham, UK

Marcin Paprzycki, Systems Research Institute Polish Academy of Sciences, Warsaw, Poland

Nicholas Caporusso, Northern Kentucky University, USA

Navanietha Rathinam, American Society for Microbiology, Rapid City, USA

Nobin Saha, University of Bedfordshire, UK

Nazia Parveen, Aligarh Muslim University, Aligarh, India

V. Neelanarayanan, VIT, Chennai, India

Neha Kaushik, DSEU, New Delhi, India

Nemi Chandra Rathore, Central University of South Bihar, Gaya, India

Nirmal Kaur, Panjab University, Chandigarh, India

N. Balasubramani, NIT, Trichy, India

Neeraj Rathore, Indira Gandhi National Tribal University (IGNTU—A Central University), Amarkantak (M.P.), India

Neelesh S. Upadhye, IIT Madras, India

Neeraj Misra, Indian Institute of Technology, Kanpur, India

Niraj Khare, Carnegie Mellon University, USA

Nitu Kumari, IIT Mandi, India

N. R. Vemuri, University of Hyderabad, Hyderabad, India

N. Anbazhagan, Alagappa University, Karaikudi, India

Neetesh Saxena, Cardiff University, UK

Nemi Chandra Rathore, Central University of South Bihar, Gaya, India

Nesibe Yalçin, Erciyes University, elikgazi/Kayseri, Turkey

Om Prakash, IIT Patna, Bihar, India

C. Oswald, National Institute of Technology Tiruchirappalli, India

Om P. Suthar, Malaviya National Institute of Technology Jaipur, Rajasthan, India

Om Prakash Yadav, NIT, Hamirpur, India

Prakash Chelladurai, Max Planck Institute, Frankfurtam, Germany

Pruthvi Balachandra Kalyandurg, Swedish University of Agricultural Sciences, Uppsala, Sweden

Pon Janani Sugumaran, NUS-National University of Singapore, Singapore

Pratheep Kumar Reddy Yaddala, Target Corporation, Minneapolis, USA

Prakash Ramalingam, Civica, Singapore, Singapore

Pushpalatha Sekar, Cognizant Technology Solutions, San Jose, USA

S. Priya Dharshini, Syntel International Pvt. Ltd., United States
P. Muthu, NIT Warangal, India
Paras Ram, NIT, Kurukshetra, Haryana, India
Patil Shrishailappa Tatyasaheb, Vishwakarma Institute of Technology, Pune, India
V. Pattabiraman, VIT, Chennai, India
Piyali Debnath, NIT Agartala, Agartala, Tripura, India
Pradip Roul, VNIT, Nagpur, India
Pramod Kumar Goyal, Bhai Parmanand DSEU Shakarpur Campus-II, Delhi, India
Prashant Giridhar Shambharkar, Delhi Technological University, New Delhi, India
Prashant Kumar Srivastava, IIT Patna, Bihar, India
Prashant R. Nair, Amrita Vishwa Vidyapeetham, Coimbatore, India
Prem Prakash Mishra, NIT, Nagaland, India
Priti Kumar Roy, Jadavpur University, Kolkata, India
Priyanka Harjule, Malaviya National Institute of Technology (MNIT), Jaipur, Rajasthan, India
Projesh Nath Choudhury, IIT Gandhinagar, Gujarat, India
Purushottam Kar, IIT Kanpur, India
Pabitra Pal, Vidyasagar University, India
Prabhat Ranjan, Central University of South Bihar, Gaya, India
Prashant Kumar Srivastava, IIT Patna, Bihar, India
Pratibhamoy Das, IIT Patna, Bihar, India
Prodipto Das, Assam University (A Central University), Silchar, India
Prof. Dr. Jana Dittmann, University of Magdeburg, Germany
Projesh Nath Choudhury, IIT Gandhinagar, Gujarat, India
P. K. Parida, Central University of Jharkhand, Ranchi, India
P. Muthukumar, IIT Kanpur, India
P. P. Murthy, Guru Ghasidas Vishwavidyalaya (A Central University), Bilaspur (Chhattisgarh), India
Pablo Berna, CUNEF Universidad, Spain
Panchatcharam Mariappan, IIT Tirupati, India
Pankaj K. Das, Tezpur University, Sonitpur, Assam, India
Pawan Kumar, Indira Gandhi National Open University Maidan Garhi, New Delhi, India
Pradip Sasmal, IIT Jodhpur, India
Prasun Ghosal, Indian Institute of Engineering Science and Technology, Shibpur, India
Praveen Kumar Gupta, National Institute of Technology Silchar, India
Predrag Stanimirovic, University of Nis, Serbia
Punam Gupta, Dr. Hari Singh Gaur Viswavidyalaya Sagar, Central University of MP, India
Puneet Sharma, IIT Jodhpur, Rajasthan, India
Patitapaban Rath, Kalinga Institute of Industrial Technology (KIIT) Deemed to be University, Odisha, Bhubaneswar, India
Pramod Kewat, IIT, Dhanbad, India

P. K. Sahoo, Birla Institute of Technology and Science, Pilani Hyderabad Campus, India

P. Venkata Suresh, Indira Gandhi National Open University, New Delhi, India

Pushkar S. Joglekar, Viswakarma Institute of Technology, Pune, India

Promila Kumar, Gargi College (University of Delhi), India

Pedro Caceres, Fort Worth Metroplex, Dallas, USA

Radhakrishna Bhat, Manipal Institute of Technology, Manipal Academy of Higher Education, Manipal, Karnataka, India

S. Rajkumar, VIT, Chennai, India

Robert Richardson, Ravensbourne University London, UK

B. R. Rakshith, MIT, Manipal, India

Rajat Kumar Pal, University of Calcutta, India

G. Rajeshkumar, Bannari Amman Institute of Technology, TN, India

Raj Nandkeolyar, NIT, Jamshedpur, India

Romi Banerjee, IIT Jodhpur, India

Ranjit Kumar Upadhyay, Indian Institute of Technology (Indian School of Mines), Dhanbad, Jharkhand, India

R. Kalyanaraman, Annamalai University, Tamil Nadu, India

Rajat Kanti Nath, Tezpur University, Assam, India

Rajendra K. Ray, Indian Institute of Technology Mandi, India

D. Ranganatha, Central University of Karnataka, India

B. V. Rathish Kumar, IIT Kanpur, India

Rabinder Kumar Prasad, D.U.I.E.T. Dibrugarh University, Assam, India

Rafikul Alam, IIT Guwahati, Guwahati, India

Rahul Kumar Chawda, Assam University, Silchar, India

Rajendra Kumar Roul, Thapar Institute of Engineering and Technology, India

Rakesh Arora, IIT (BHU), Varanasi, India

Ranbir Sanasam, Indian Institute of Technology Guwahati, India

Ratikanta Behera, IISc Bangalore, India

Rohit Kumar Mishra, IIT Gandhinagar, Gujarat, India

Rupam Barman, Indian Institute of Technology Guwahati, Assam, India

Rifat Colak, Firat University, Türkiye

Reshma Rastogi, South Asian University, New Delhi, India

Ravi Subban, Pondicherry University, Puducherry, India

Ravi Kanth Asv, National Institute of Technology Kurukshetra, India

R. Eswari, NIT Trichy, India

R. Radha, University of Hyderabad, Hyderabad, India

R. Jagadeesh Kannan, Vellore Institute of Technology | Chennai Campus, Chennai, India

R. Kalyanaraman, Annamalai University, Tamil Nadu, India

R. Meher, S. V. National Institute of Technology, Surat, Gujarat, India

R. Suganya, VIT Chennai, India

Rabinder Kumar Prasad, D.U.I.E.T. Dibrugarh University, Assam, India

Rahul Kumar Chawda, Assam University, Silchar, India

Raj Kamal Maurya, SVNIT, Surat, Gujarat, India

Rajat Kanti Nath, Tezpur University, Assam, India
Rajat Tripathi, NIT, Jamshedpur, Jharkhand, India
Rajendra Kumar Roul, Thapar Institute of Engineering and Technology, India
Rajesh Ingle, International Institute of Information Technology, Naya Raipur, (IIIT NR), Chhattisgarh, India
Raksha Pandey, Guru Ghasidas Vishwavidyalaya, Bilaspur, India
Ramesh Kumar Vats, NIT Hamirpur, India
Ramesh Ragala, VIT Chennai, India
Ranjan Kumar Jana, Sardar Vallabhbhai National Institute of Technology (SVNIT) Surat, Gujarat, India
Ranjit Kumar Upadhyay, Indian Institute of Technology (Indian School of Mines), Dhanbad, Jharkhand, India
B. V. Rathish Kumar, IIT Kanpur, India
Reshma Rastogi (nee Khemchandani), South Asian University, New Delhi, India
Rifaqat Ali, NIT, Hamirpur, India
Ritu Agarwal, Malaviya National Institute of Technology Jaipur, Rajasthan, India
Rathina Kumar, University of Virginia, Charlottesville, USA
Raghunath Vel, University of Wolverhampton, UK
Rajaguru Paramasamy, H-E-B, Sr. PeopleSoft, San Antonio, USA
Ramprasad Renganathan, Omnitracs, North Atlanta, USA
Radhakrishnan Seenivasan, NTT DATA Americas, Pittsburgh, USA
Rajkumar Kathiresan, Cognizant Technology, Bentonville, USA
Revathi Balasubramanian, DFKI, Kaiserslautern, Germany
Raja Sekar, Cognizant Technology Solution, Toronto, Canada
Ram Bharadwaj, HCL Technologies, Ottawa, Canada
Rakesh Prabhakar, Cognizant, Winnipeg, Canada
Raja Saravanesh, Cognizant, Owings Mills, USA
Rajesh Kannan Karuppiah, Tata Consultancy Services, Jesus Martin, Mexico
Raja Lingam, Budapest University of Technology and Economics, Hungary
K. R. Renjith, Servian, Sydney, Australia
Ramsundar Kandasamy, Ericsson R&D, Germany
Raja Prabhu, Anya Consultancy Services, UK
B. Rushi Kumar, VIT Vellore, Tamil Nadu, India
Sri Padmavati. B, University of Hyderabad, Telangana, India
Sri Balaji Ponraj, The University of Sydney, Australia
Subathra Kannan, Biocon, Trichy, France
Swati Sharma, North Dakota State University, Fargo, USA
Satheesh Kumar, Nordic BioAnalysis AB, Stockholm, Sweden
Sannasi Nehru, Deloitte Consulting, Washington, USA
Sukumar Subburayan, CISCO, USA
Smitha Samuel, Sabre Inc. Irving, USA
Sangram Ray, National Institute of Technology Sikkim, India
Selvam Adaikkappan, Vistex, Chicago, USA
Saravanakumar Jagadeesan, Deloitte Consulting US, Austin, USA
Sujith Vijayakumar, Infosys Technologies Ltd., Dallas, USA

Sunil Ramkumar, Tata Consultancy Services, London, UK
Sadheesh Radhakrishnan, S-Cube Solutions Ltd. Location: London, UK
N. Shyam Sundar, Motorola Solutions, Plantation, USA
Siddharth Gopinath, Gainwell Technologies, Columbus, USA
Saravanan Sukumar, Cognizant Technology Solutions, Charlotte, USA
Saranya Kamarajan, TCS, USA
Sundar Ravanan, eHarmony, United States
Sowbhagya Lakshminarayanan, Greatness Packagers, Canada
Shanmuganathan, Tata Consultancy Services, United States
Suresh Raja, Tata Consultancy Services, United States
S. A. M. Rizvi, Jamia Millia Islamia, New Delhi, India
S. Amutha, Alagappa University, Karaikudi, India
S. Nithya Roopa, KIT, Coimbatore, India
S. R. Balasundaram, NIT, Trichy, India
Sakthi Prasad, National Institute of Technology, Arunachal Pradesh, India
Sam Johnson, NIT Karnataka, Suratkal, India
Sandeep Shinde, Vishwakarma Institute of Technology, Pune, India
Sanjeev Kumar, NSUT West Campus Jaffarpur Delhi, India
Saurabh Kumar Katiyar, NIT, Jalandar, Punjab, India
Shachi Sharma, South Asian University, New Delhi, India
Shafiqul Abidin, Aligarh Muslim University Aligarh, India
Shakir Ali, Aligarh Muslim University, Aligarh, India
Shanmugam Dhinakaran, Indian Institute of Technology Indore, India
Shraddha S. Suratkar, VJTI-Mumbai, India
Siddhartha Pratim Chakrabarty, Indian Institute of Technology Guwahati, Assam, India
Subrata Bera, National Institute of Technology Silchar, Assam, India
Subuhi Khan, Aligarh Muslim University, India
Sujoy Bhore, IIT Bombay, Mumbai, India
Sumit Kumar Debnath, NIT, Jamshedpur, Jharkhand, India
Sumit Nagpal, University of Delhi, India
Suraiya Jabin, Jamia Millia Islamia (Central University), New Delhi, India
Surendar Ontela, NIT, Mizoram, India
Susantha Maity, National Institute of Technology, Arunachal Pradesh, India
Sushil Kumar, S. V. National Institute of Technology Surat, Gujarat, India
T. R. Swapna, Amrita Vishwa Vidyapeetham, Coimbatore, India
S. Ponnusamy, IIT Madras, India
Saibal Pal, DRDO, New Delhi, India
Sangram Ray, National Institute of Technology Sikkim, India
Sanjay Mohanty, VIT Vellore, India
Santanu Sarkar, IIT Madras, India
Sarita Ojha, Indian Institute of Engineering Science and Technology Shibpur, West Bengal, India
Satrajit Ghosh, Aarhus University, Denmark
Sedat Akleylek, Ondokuz Mayis University, Turkey

P. Shaini, Central University of Kerala, Kerala, India
Sharanjeet Dhawan, NIIT University, Rajasthan, India
Sharmistha Adhikari, NIT Sikkim, India
Sk Hafizul Islam, IIIT Kalyani, India
Sokratis Katsikas, Norwegian University of Science and Technology, Norway
Sourav Mandal, XIM University Bhubaneswar, Odisha, India
Subhas Barman, Jalpaiguri Government Engineering College, West Bengal, India
Subhasis Dasgupta, University of California, San Diego, USA
Suprio Bhar, IIT Kanpur, India
Syed Abbas, IIT Mandi, India
Sandeep Singh Rawat, IGNOU, New Delhi, India
A. Sathishkumar, IIT Madras, India
Sanyasiraju, IIT Madras, India
S. P. Tiwari, Indian Institute of Technology (Indian School of Mines), Dhanbad, Jharkhand, India
Srinivas Kumar Vasana, Indian Institute of Technology Delhi, India
Srinivasa Rao Pentyala, Indian Institute of Technology (ISM) Dhanbad, India
Suchandan Kayal, National Institute of Technology Rourkela, Odisha, India
Sanjeev Singh, Indian Institute of Technology, Indore, MP, India
K. C. Srikantaiah, SJB Institute of Technology, Bengaluru, Karnataka, India
Sangita Jha, NIT Rourkela, India
Sairam Kaliraj, IIT Ropar, Punjab, India
Sreenivasulu Ballem, Central University of Karnataka Kalaburagi, India
Sanjay Kumar, Central University of Jammu, India
S. K. Pandey, IIT (BHU, Varanasi), India
S. P. Tiwari, Indian Institute of Technology (Indian School of Mines), Dhanbad, Jharkhand, India
S. K. V. Jayakumar, Pondicherry University, Pudhucherry, India
Sabyasachi Dutta, University of Calgary, Canada
Sabyasachi Pani, IIT Bhubaneswar, India
Sahana Prasad, BITS Pilani, Rajasthan, India
Saifur Rahman, Rajiv Gandhi University Doimukh, India
Saiyed Umer, ISI Kolkata, India
Saminathan Ponnusamy, IIT Madras, India
Santanu Saha Ray, National Institute of Technology Rourkela, Odisha, India
Sartaj Ul Hasan, Indian Institute of Technology Jammu, India
Sasmita Barik, IIT Bhubaneswar, India
Sujit Das, NIT Warangal, India
A. Sathish Kumar, IIT Madras, India
Satyanarayana Engu, NIT, Warangal, Andhra Pradesh, India
Sedat Akleylek, Ondokuz Mayis University, Turkey
Shailesh Kumar Tiwari, Indian Institute of Technology Patna, Bihar, India
Shibesh Kumar Jas Pacif, VIT, Vellore, India
Shripad M. Garge, Indian Institute of Technology, Mumbai, India

Shyamalendu Kandar, Indian Institute of Engineering Science and Technology, Shibpur, India

Siddhartha Pratim Chakrabarty, Indian Institute of Technology Guwahati, Assam, India

Siuli Mukhopadhyay, Indian Institute of Technology Bombay, Mumbai, India

Sivaram Ambikasaran, IIT Madras, India

Somesh Kumar, Indian Institute of Technology Kharagpur, India

Somnath Dey, Indian Institute of Technology Indore, India

Subinoy Chakraborty, Jadavpur University, Kolkata, West Bengal, India

Srinivasa Rao Kola, NIT, Karnataka, Suratkal, India

Subit Kumar Jain, NIT, Hamirpur, India

Srinivasu Bodapati, IIT Mandi, India

Subir Das, Indian Institute of Technology (BHU), Varanasi, India

Sudesh Rani, Punjab Engineering College, Deemed to be University, Chandigarh, India

Sudipta Majumder, Dibrugarh University Institute of Engineering and Technology (DUIET), Assam, India

Sujata Pal, IIT Ropar, India

K. Sumesh, IIT Madras, Chennai, India

S. Bose, College of Engineering, Anna University, India

Sujoy Bhore, IIT Bombay, India

Satya Bagchi, NIT Durgapur, India

Swaleha Zubair, Aligarh Muslim University, India

S. Gandhiya Vendhan, Bharathiar University, Coimbatore, India

Sujit Das, NIT, Warangal, India

Shafik, Nanjing University of Information Science and Technology, China

Scott Baldridge, Baton Rouge, LA, USA

Tarak Gaber, The University of Salford, UK

Terry Kaufman, Institute for Mathematics and Computer Science, Fort Lauderdale, FL, USA

Triloki Nath, Dr. Hari Singh Gour Vishwavidyalaya, Sagar, MP, India

Taqseer Khan, Jamia Millia Islamia, Jamia Nagar, New Delhi, India

Tanmoy Maitra, KIIT University, Bhubaneswar, India

Tapas Chatterjee, IIT Ropar, Punjab, India

Tarun Yadav, Defence Research and Development Organisation, New Delhi, India

Tingwen Huang, Texas A&M University, USA

Tuhina Mukherjee, Indian Institute of Technology Jodhpur, India

T. Subbulakshmi, VIT, Chennai, India

Tamal Pramanick, NIT Calicut, Kozhikode, Kerala, India

Tanweer Jalal, NIT, Srinagar, India

Tarni Mnadal, NIT, Jamshedpur, Jharkhand, India

Thomas George, Missing Link Technologies Ltd., Moncton, Canada

Thirumaran Pathakkam Mannai, Elavon, Inc. Knoxville, USA

Ujwal Warbhe, NIT, Srinagar, India

Usha Rani, Sri Venkateswara University, Tirupati, Andhra Pradesh, India

Uaday Singh, Indian Institute of Technology Roorkee, Roorkee-247667 (Uttarakhand), India

Ushnish Sarkar, Netaji Subhas Open University, India

Utpal Roy, Siksha-Bhavana Visva-Bharati Santiniketan, Birbhum, WB, India

Udayan Prajapati, St. Xavier's College, Navrangpura, Ahmedabad, India

Vilem Novak, University of Ostrava, Czech Republic

Vinod Kumar, PGDAV College, University of Delhi, Nehru Nagar, New Delhi, India

V. Balakumar, National Institute of Technology Puducherry, Karaikal, India

Vishnu Narayan Mishra, Indira Gandhi National Tribal University, Madhya Pradesh, India

V. V. Subrahmanyam, IGNOU, New Delhi, India

V. Shanthi, NIT Trichy, Tamil Nadu, India

V. D. Ambeth Kumar, Mizoram University, Mizoram, India

Vipindev Adat Vasudevan, Massachusetts Institute of Technology, Cambridge, USA

V. Shanthi, NIT Trichy, Tamil Nadu, India

Vaibhav Dhore, VJTI-Mumbai, India

R. Vedhapriyavadhana, VIT, Chennai, India

Vijayakumar Ramakrishnan, NIT, Calicut, Kerala, India

K. V. Vijayashree, Anna University, Chennai, India

Vinay Singh, NIT Mizoram, India

Vipindev Adat Vasudevan, Massachusetts Institute of Technology, Cambridge, India

Venkatesh Babu Nattamai Balakrishnan, Diconium Digital Solutions, Berlin, Germany

Venguideshe (Venkat) Lakshminarayanan, ServiceNow, Santa Clara, USA

Valli Elangovan, Tata Consultancy Services, Edinburgh, UK

Vijay Shankar, Cognizant, New York, USA

Venkat Sundaram, e-Business International, United States

P. L. Valliappan, ADP Technologies, United States

Venkatesh Srinivason, Tata Consultancy Services Limited, United States

Vasos Pavlika, University College London, UK

Yuvaraj Nagarajan, VIT Infotech, San Ramon, USA

Yamuna Venkates, IStream Jobs, France

Y. D. Sharma, NIT Hamirpur, India

Yegnanarayanan Venkataraman, Kalasalingam Academy of Research and Education, Tamil Nadu, India

Zhiqiang Lin, The Ohio State University, USA

Message from General Chairs

It is really a happy moment for all of us to greet you all at the ICMC 2024, the 10th edition of the highly reputed annual International Conference on Mathematics and Computing. This year, ICMC was organized at the main campus of KALASALINGAM ACADEMY OF RESEARCH AND EDUCATION-KARE located at a small village in the interior of South Tamil Nadu, Krishnankoil-626126, Tamil Nadu, India. The event was held during January 04–07, 2024 followed by a Pre-Conference Symposium on Advanced Mathematical Methods during 02–03, January 2024. It was organized by the Department of Mathematics of School of Advanced Sciences jointly with School of Computing. We are happy to record the fact that the Ramanujan Mathematical Society (RMS), Cryptology Research Society India (CRSI) and Society for Electronics Transactions and Security (SETS) have participated as joint organizers with KARE. Due to the collective effort of the organizers, we have successfully brought on board 610 eminent personalities as Technical Programme Committee members from all over the world and eighteen eminent speakers from across the globe. It really paved the way for high-quality academic and technical exchange of thought processes between likeminded delegates. As usual, this time also ICMC has created a great impact among the participating countries, and everyone have thoroughly enjoyed the academic ambience and the location with the scenic beauty of the southern parts of the Western Ghats of Tamil Nadu. The central theme and topics on mixed areas of Mathematics and Computing created a strong impact. Original research articles published from this 10th edition of ICMC stand as a testimony for the hard work of researchers. We are delighted to record a wonderful fact that total 40 papers have been considered for publication in the conference proceeding out of 282 submitted papers through the hard peer-review process by the TPC members. We are much grateful to the following invited speakers for their graceful acceptance and also for having delivered their best of the best speech at the conference. The honourable speakers are Elisa Bertino (Purdue University, USA), Bavani Thuraisingham (University of Texas, USA), Muriel Medard (MIT, USA), Ramamohanarao Kottagiri (Universiyty of Melbourne, Australia), Mohammad S. Obaidat (University of Texas-Permian Basin, USA), Sedat Akleylek (Ondokuz Mayis

University, Samsun, Turkey), Ekrem Savas (Usak University, Turkey), Bryan Frey-berg (University of Minnesota, USA), Mark Sepanski (Baylor University, Waco Texas, USA), Clare D. Cruz (Chennai Mathematical Institute, India), Jaya Iyer (The Institute of Mathematical Sciences, India), Arvind Ayyar (Indian Institute of Science, India), R. K. Sharma (IIT Delhi, India), Tanmoy Som (IIT (BHU) Varanasi India), Tanmay Basak (IIT Madras, India), Madhumangal Pal (Vidyasagar University, India), A. K. B. Chand (IIT Madras, India) and Vishnu Pendyala (San Jose State University, USA).

We express our sincere gratitude to all the programme and organizing committee members for their fantastic review process and other works. We would also like to thank "Illaya Vallal" Dr. Sridharan, Chancellor, and his team at Kalasalingam Academy of Research and Education for their support and excellent infrastructure. **We also thank the sponsors, The Defence Research and Development Organisation (DRDO) and The National Board for Higher Mathematics (NBHM), Government of India, for financial support**. We also appreciate all conference participants for making ICMC a memorable one.

General Chairs

P. K. Saxena, Scientific Consultant—Cyber Security and Former Director, SAG, DRDO, India
P. D. Srivastava, Indian Institute of Technology Bhilai, Raipur, India

Message from Programme Chairs

We are used to the practice of allotting the task of organizing the series of ICMC conferences to good institutions spread across the country. This time, we are much pleased to award the task of organizing the 10th edition of ICMC 2024 to the KALASALINGAM ACADEMY OF RESEARCH AND EDUCATION-KARE located at a small village in the interior of South Tamil Nadu, Krishnankoil-626126, Tamil Nadu, India. The event was held during January 04–07, 2024, at the KARE University main campus followed by a Pre-Conference Symposium on Advanced Mathematical Methods during 02–03, January 2024. It was organized by the Department of Mathematics of School of Advanced Sciences jointly with School of Computing. The aim is to provide a common platform for researchers and experts from both Mathematics and Computing to meet, exchange ideas and learn the recent happenings in the respective fields. The speakers are carefully selected to give the best experience to the prospective audience and young researchers. Eighteen speakers from India and abroad delivered their talks, and some of them gracefully accepted to act as session chairs. The response was really wonderful this time to our call for paper through Easychair. We received 282 papers for the conference. There was an overwhelming response from eminent people of top-class institutions from all over the world. We are fortunate to have 610 members in our Technical Programme Committee. We followed triple blind-review process and carefully selected only 40 articles for publication in the conference proceedings published by the Springer series: *Lecture Notes in Networks and Systems*.

The 10th edition of ICMC 2024 has earned a good repute across countries like India, the USA, Canada, Australia, Japan, France, Germany, China, Indonesia, Turkey, the UAE and Nigeria. The delegates from these countries participated and exchanged their thoughts in a variety of areas of pure mathematics, applied mathematics and computing. We are very thankful to the chief patron, patrons, co-patrons, general chairs, programme chairs, organizing chair, speakers, participants, referees, organizers, sponsors and funding agencies for their support and help. Our special thanks to Ramanujan Mathematical Society, Cryptology Research Society of India and Society for Electronic Transactions and Security for coming forward to organize this event jointly with us. Last but not the least, we record here our soulful thanks to all the volunteers and the workers at the grassroots level who worked tirelessly to make this event a memorable one. A well-planned teamwork was the real reason behind the success of this conference.

New Jersey, USA   Jaideep Vaidya
Chennai, India   S. Ponnusamy
Columbus, USA   Zhiqiang Lin
Baltimore, USA   Karuna Pande Joshi
Kolkata, India   Debasis Giri

Preface

Mathematics reveals hidden patterns that help us to understand the world around us. Now, much more than arithmetic and geometry, mathematics today is a diverse discipline that deals with data, measurements and observations from science, with inference, deduction and proof; and with mathematical models of natural phenomena, of human behaviour and of social systems. Man is a social animal, and human life depends upon the cooperation of each other. Group work helps social skills. The ability to work together on tasks with others can build various social skills. The importance of Mathematics and Computing as a tool for science and technology is continually increasing. While science and technology have become so pervasive, mathematics and computing education have continued to dominate the curriculum and remain a key subject area requirement in the higher education and the employment sector. Mathematics and Computing are being applied to agriculture, ecology, epidemiology, tumour and cardiac modelling, DNA sequencing and gene technology. They are used to manufacture medical devices and diagnostics, and sensor technology

The ICMC conference series began its service since 2013 at Haldia Institute of Technology, India. ICMC was further conducted by many reputed institutes such as IIT (BHU), KIIT, Bhubaneswar and Sikkim University, India. The 10th edition of the ICMC series has been conducted in offline mode. ICMC 2024 will provide a wonderful opportunity for both young and seasoned scientists to meet each other to share new ideas and to provide a space for researchers from both academic and industry to present their original work in the area of Computational Applied Mathematics that comprises topics such as Operations Research, Numerical Analysis, Computational Fluid Mechanics, Soft Computing, Cryptology and Security Analysis, Image Processing, Big Data, Cloud Computing, Data Analytics, IoT, Pervasive Computing, Computational Graph Theory and other emerging areas of research.

The 10th Internal Conference on Mathematics and Computing (ICMC 2024) was organized by the Department of Mathematics of School of Advanced Sciences and School of Computing, of Kalasalingam Academy of Research and Education-KARE, Krishnankoil-626126, Tamil Nadu, India. The Ramanujan Mathematical Society (RMS), Cryptology Research Society India (CRSI) and Society for

Electronics Transactions and Security (SETS) also participated as joint organizers of ICMC 2024.

Original research articles published from this 10th edition of ICMC stand as a testimony to the hard work of researchers. We are delighted to record a wonderful fact that total 40 papers have been considered for publication in the conference proceeding out of 282 submitted papers through the hard peer-review process by the TPC members. We are much grateful to the following invited speakers for their graceful acceptance and also for having delivered their best of the best speech at the conference. The honourable speakers are Elisa Bertino, Purdue University, USA, Bavani Thuraisingham, University of Texas, USA, Muriel Medard, MIT, USA, Ramamohanarao Kottagiri, University of Melbourne, Australia, Mohammad S. Obaidat, USA, Sedat Akleylek, Ondokuz Mayis University, Samsun, Turkey, Ekrem Savas, Usak University, Turkey, Bryan Freyberg, University of Minnesota, USA, Mark Sepanski, Baylor University, Waco Texas, USA, Clare D. Cruz, CMI, India, Jaya Iyer, IMSC, India, Arvind Ayyar, ISSC, India, R. K. Sharma, IIT Delhi, Tanmoy Som, IIT (BHU), India, Tanmay Basak, IIT Madras, Madhumangal Pal, Vidyasagar University, India, A. K. B. Chand IIT Madras, India and Vishnu Pendyala, San Jose State University, USA.

A unique feature of this book series is that quality assurance is facilitated by a rigorous selection process through the Easychair submission method. Six hundred and ten eminent people from all over the world have participated as TPC members to select the submitted papers. This book contains carefully papers of high-quality researchers in two volumes of 20 papers each as chapters. These two volumes (Volume I and Volume II) speak an exhaustive literature survey, the bottlenecks and advancements in several areas of mathematics and computing that happened in this decade. The excellent coverage of these two volumes is at a higher level to fulfil the global requirements of mathematics, computing and their applications in science and engineering. The audience of this book are mainly researcher scholars, scientists, mathematicians and people from industry.

As Volume Editors of these two volumes, we gratefully acknowledge all the administrative authorities of Kalasalingam Academy of Research and Education-KARE, for their encouragement and support. We also express our appreciation to all the faculty members and research scholars of the department of mathematics and School of Computing of Kalasalingam Academy of Research and Education-KARE. We specially thank the Chief Patron, "Illaya vallal" Dr. Sridharan, the Chancellor of KARE, General Chairs, programme chairs and all the members of the organizing committee of ICMC 2024 who contributed as one unit by dedicating their time to make the conference a memorable one. We sincerely acknowledge all the referees for their valuable time in reviewing the manuscripts and for carefully picking the best original research articles for publication. We also record our special mention to the sponsors, The Defence Research and Development Organisation (DRDO) and The National Board for Higher Mathematics (NBHM), Government of India, for liberal financial grant. Finally, we are very glad to Springer (Lecture Notes in Networks and Systems) for their encouragement and guidance towards the publication of the proceedings of the conference as two volumes.

Kolkata, India   Debasis Giri
New Jersey, USA   Jaideep Vaidya
Chennai, India   S. Ponnusamy
Columbus, USA   Zhiqiang Lin
Baltimore, USA   Karuna Pande Joshi
Srivilliputhur, India   V. Yegnanarayanan

Contents

Editors and Contributors

About the Editors

Debasis Giri is at present Associate Professor in the Department of Information Technology of Maulana Abul Kalam Azad University of Technology (Formerly known as West Bengal University of Technology), West Bengal, India prior to Professor (in Computer Science and Engineering) and Dean (in School of Electronics, Computer Science and Informatics) of Haldia Institute of Technology, Haldia, India. He did his masters (M.Tech. and M.Sc.) both from IIT Kharagpur, India, and also completed his Ph.D. from IIT Kharagpur, India. He is tenth all India rank holder in Graduate Aptitude Test in Engineering in 1999. He has published more than 100 papers in international journal/conference. His current research interests include Cryptography, Information Security, Blockchain Technology, E-commerce Security and Design and Analysis of Algorithms. He is Editorial Board Member and Reviewer of many International Journals. He is also Program Committee Member of International Conferences. He is a life member of Cryptology Research Society of India, Computer Society of India, the International Society for Analysis, its Applications and Computation (ISAAC) and IEEE annual member.

Jaideep Vaidya is a Distinguished Professor of Computer Information Systems at Rutgers University and the Director of the Rutgers Institute for Data Science, Learning, and Applications. He received the B.E. degree in Computer Engineering from the University of Mumbai, the M.S. and Ph.D. degree in Computer Science from Purdue University. His general area of research is in security, privacy, data mining, and data management. He has published over 200 technical papers in peer-reviewed journals and conference proceedings, and has received several best paper awards from the premier conferences in data mining, databases, digital government, security, and informatics. He is an IEEE and AAAS Fellow as well as an ACM Distinguished Scientist. He served as the Editor in Chief of the IEEE Transactions on Dependable and Secure Computing.

S. Ponnusamy is currently the Chair Professor at IIT Madras, and the President of the Ramanujan Mathematical Society, India. His research interest includes complex analysis, special functions, and functions spaces. He served five years as a Head of the Indian Statistical Institute, Chennai Centre. He is the Chief Editor of the Journal of Analysis and serves as a Editorial member for many peer reviewed international journals. He has written five text books and has edited several volumes, and international conference proceedings. He has solved several long standing open problems and conjectures, and published more than 300 research articles in reputed international journals. He has been a Visiting Professor to a number of universities in abroad (e.g. Hengyang Normal University, Hunan First Normal University and Hunan Normal University; Kazan Federal University and Petrozavodsk State University; University Sains Malaysia; University of Aalto, University of Turku, and University of Helsinki; University of South Australia; Texas Tech University). Currently, he is also a Leader of the group on the geometric theory of functions at the Laboratory "Multidimensional Approximation and Applications" of the Lomonosov Moscow State University, Moscow Center for Fundamental and Applied Mathematics, Moscow, Russia. He is also a Chair Professor "Furong Scholars Award Program", of Hunan First Normal University, China.

Zhiqiang Lin is a Distinguished Professor of Engineering, and the director of Institute for Cybersecurity and Digital Trust (ICDT) at The Ohio State University. His research interests center around systems and software security, with a key focus on (1) developing automated binary analysis techniques for vulnerability discovery and malware analysis, (2) hardening the systems and software from binary code rewriting, virtualization, and trusted execution environment, and (3) the applications of these techniques in Mobile, IoT, Bluetooth, and Connected and Autonomous Vehicles. He has published over 140 papers, many of which appeared in the top venues in cybersecurity. He is an ACM Distinguished Member, a recipient of Harrison Faculty Award for Excellence in Engineering Education, NSF CAREER award, AFOSR Young Investigator award, and Outstanding Faculty Teaching Award. He received his Ph.D. in Computer Science from Purdue University.

Karuna Pande Joshi is an Associate Professor of Information Systems at the University of Maryland, Baltimore County (UMBC). She is the UMBC Director for the Center of Accelerated Real Time Analytics (CARTA) and the Director of the Knowledge, Analytics, Cognitive, and Cloud (KnACC) Lab. She is also the Undergraduate Program Director of the Business Technology Administration Program. Her primary research focus is Data Science, Legal Text Analytics, Cloud Computing, and Health IT. She has published over 90 technical papers in peer-reviewed journals and conference proceedings. Dr. Joshihas been awarded research grants by NSF, ONR, DoD, NIH, Cisco, and GE Research. She received her M.S. and Ph.D. in Computer Science from UMBC, where she was twice awarded the IBM Ph.D. Fellowship. She did her Bachelor of Engineering (Computers) from the University of Mumbai. Dr. Joshi has also worked for over 15 years in the Industry, including as a Senior Information Management Officer at the International Monetary Fund for nearly a decade.

V. Yegnanarayanan is a Senior Professor of Mathematics at Kalasalingam Academy of Research and Education, Tamilnadu, India. His Erdos Number is three. He has authored 215 Research papers in reputed refereed journals and Conferences and published eight Patents in India. So far he has produced six Ph.D's. He was elected as a Senior Member of IEEE for his meritorious contributions to the cause of technical education and research in 2012. He has won the prestigious Sentinel of Science Award by Publons, UK in the year 2016 for his contributions to Review work of research papers submitted to journals in Mathematics. He is a life member and affiliate members of various professional societies like AMS, SIAM, RMS, IMS, ISTE. He has successfully completed funded research projects and organized SDP's funded by AICTE, Conferences sponsored by Tamilnadu State Council for Science and Technology and delivered a number of invited talks in India and Abroad.

Contributors

Mohammad Alakhrass University of Sharjah, Sharjah, UAE

Siddhartha P. Chakrabarty Department of Mathematics, Indian Institute of Technology Guwahati, Guwahati, India

Naveen Chandra Bhagat Department of Mathematics, Central University of Jharkhand, Ranchi, India

J. Christy Roja Department of Mathematics, St. Joseph's College, Affiliated to Bharathidasan University, Thiruchirappalli, Tamil Nadu, India

Rifat Çolak Department of Mathematics, Firat University, Elazığ, Turkey

Levin Dabhi Intel Corporation, Bangalore, India

A. K. Das Indian Statistical Institute, Kolkata, India

Soumen De Department of Applied Mathematics, University of Calcutta, Kolkata, India

R. Deb Jadavpur University, Kolkata, India

Lavanya Elluri Subhani Department of Computer Information Systems, Texas A&M University-Central Texas, Killeen, TX, USA

E. A. Gopalakrishnan Amrita School of Artificial Intelligence, Amrita Vishwa Vidyapeetham, Banglore, India

Priya Gulati Department of Mathematics, Indian Institute of Technology Guwahati, Guwahati, India

Aadi Gupta Department of Mathematics, Indian Institute of Technology Guwahati, Guwahati, India

Samar Idris Graduate School of Sciences, Firat University, Elazığ, Turkey; Graduate School of Sciences, Firat University, Elazığ, Turkey

M. Jayasudha IFET College of Engineering, Villupuram, India

V. Jayasudhan IFET College of Engineering, Villupuram, India

J. Jenifa Department of Mathematics, St. Joseph's College, Affiliated to Bharathidasan University, Thiruchirappalli, Tamil Nadu, India

Karuna Pande Joshi Department of Information Systems, University of Maryland, Baltimore County, MD, USA

T. Kalaiselvi Department of Mathematics, Kalasalingam Academy of Research and Education, Krishnankoil, Tamilnadu, India

R. Karthika Department of Mathematics, School of Mathematics and Computer Sciences, Central University of Tamil Nadu, Thiruvarur, India

Harsh Kedia Intel Corporation, Bangalore, India

Samhita Konduri Palo Alto High School, Palo Alto, CA, USA

Senthil Kumaran R IFET College of Engineering, Villupuram, India; SRM Institute of Science and Technology, KTR Campus, Chengalpattu, India

Chandni Kumari GLA University, Mathura, India

Sapan Kumar Nayak Department of Mathematics, Central University of Jharkhand, Ranchi, India

V. Manikandan Department of Mathematics, Manonmaniam Sundaranar University, Tirunelveli, Tamil Nadu, India

Babita Mehta Department of Mathematics, Central University of Jharkhand, Ranchi, India

Archie Mittal Intel Corporation, Bangalore, India

S. Monikandan Department of Mathematics, Manonmaniam Sundaranar University, Tirunelveli, Tamil Nadu, India

M. Nimal Madhu Amrita School of Artificial Intelligence, Amrita Vishwa Vidyapeetham, Coimbatore, India

Shwet Nisha Department of Mathematics, Gaya College of Engineering, Gaya, India

Phani Kumar Nyshadham Intel Corporation, Bangalore, India

Hidenori Ogata Department of Computer and Network Engineering, Graduate School of Informatics and Engineering, The University of Electro-Communications, Chofu, Japan

P. K. Parida Department of Mathematics, Central University of Jharkhand, Ranchi, India

Kriti V. Pendyala University Preparatory Academy, San Jose, CA, USA

Vishnu S. Pendyala Department of Applied Data Science, San Jose State University, San Jose, CA, USA

Parvathi Pradeep Amrita School of Artificial Intelligence, Amrita Vishwa Vidyapeetham, Coimbatore, India

B. Premjith Amrita School of Artificial Intelligence, Amrita Vishwa Vidyapeetham, Coimbatore, India

S. B. Ramkumar Department of Mathematics, School of Mathematics and Computer Sciences, Central University of Tamil Nadu, Thiruvarur, India

P. Ranjitha IFET College of Engineering, Villupuram, India

V. Renukadevi Department of Mathematics, School of Mathematics and Computer Sciences, Central University of Tamil Nadu, Thiruvarur, India

Biman Sarkar Center of Excellence for Ocean Engineering, National Taiwan Ocean University, Keelung, Taiwan

Ekrem Savas Department of Mathematics, Uşak University, Uşak, Turkey

Priya Sharma Department of Physics, Swami Vivekananda University, Barrackpore, Kolkata, India;
Department of Applied Mathematics, University of Calcutta, Kolkata, India

Yegnanarayanan Venkatraman Department of Mathematics, Kalasalingam Academy of Research and Education, Krishnankoil, Tamilnadu, India

Redwan Walid Department of Information Systems, University of Maryland, Baltimore County, MD, USA

S. Yuvaraj IFET College of Engineering, Villupuram, India

Analysis of Oblique Wave Scattering by a Thick Bottom-Standing Barrier Placed in Between a Pair of Thin Partially Immersed Barriers

Priya Sharma, Biman Sarkar, and Soumen De

Abstract The study investigates the interaction of oblique water waves with a configuration consisting of a pair of partially immersed thin vertical barriers on the two sides of a bottom-standing rectangular thick barrier. The eigenfunction expansion method is employed to analyze the system, leading to weakly singular Fredholm-type integral equations. Singularities near the edges of the barriers are addressed using Chebyshev and ultra-spherical Gegenbauer polynomials as basis functions. Numerical estimations of reflection and transmission coefficients are presented, demonstrating excellent agreement with existing literature and validating the theory's reliability and applicability in practical wave interaction scenarios.

Keywords Thin barriers · Bottom-standing thick barrier · Fredholm integral equations · Half and one-third singularities · Multi-term Galerkin technique · Reflection and transmission coefficients

1 Introduction

Integral equations are widely recognized and extensively utilized as a versatile tool for solving boundary value problems in various fields, including crack mechanics, acoustics, electromagnetism, water wave interaction, and more. They play a crucial role in diverse areas such as combustion, heat transfer, reaction chemistry, shallow water problems, and plasma physics. The method of matched asymptotic expan-

P. Sharma
Department of Physics, Swami Vivekananda University, Barrackpore, Kolkata 700121, India

B. Sarkar
Center of Excellence for Ocean Engineering, National Taiwan Ocean University, Keelung 202301, Taiwan

P. Sharma · S. De (✉)
Department of Applied Mathematics, University of Calcutta, 92, A.P.C. Road, Kolkata 700009, India
e-mail: soumenisi@gmail.com

© The Author(s), under exclusive license to Springer Nature Singapore Pte Ltd. 2024
D. Giri et al. (eds.), *Proceedings of the Tenth International Conference on Mathematics and Computing*, Lecture Notes in Networks and Systems 964,
https://doi.org/10.1007/978-981-97-2066-8_1

1

sion, initially introduced by Tuck [1], has been extended and applied by Packham
and Williams [2] and Mandal [3] to investigate wave transmission through narrow
gaps in vertical walls. These studies utilize integral equations and Havelock's expansion of wave potential to analyze the problem, building upon Tuck's original work
and providing additional insights and advancements in the understanding of wave
transmission phenomena.

In the field of wave scattering by breakwaters, researchers have extensively studied the interaction of water waves with thick barriers of different configurations.
Mei and Black [4] laid the groundwork by investigating wave scattering involving
thick vertical barriers with rectangular cross-sections. Guiney et al. [5] expanded
on this work by considering the effect of thickness in vertical walls, using the conformal mapping method. Kanoria et al. [6] and Mandal and Kanoria [7] explored
wave scattering scenarios for rectangular thick barriers with normal and oblique
incident waves. More recent studies by Paul and De [8] and Sasmal and De [9] have
investigated wave scattering by thick barriers with various geometric configurations,
considering additional factors such as thin ice-sheets or surface tension.

In recent decades, numerous analytical studies have focused on the scattering of
linear water waves by vertical thin rigid barriers. Several authors, such as Ursell [10]
and Levine and Rodemich [11], and the literature cited therein have made significant
contributions in investigating and explicitly solving scattering problems involving
normal incidence of waves on vertical barriers in deep water.

In this paper, we address a complex wave scattering problem involving a combination of thick and thin barriers, encompassing singularities at the edges of both types of
barriers. By transforming the problem into a system of Fredholm-type integral equations, we derive unknown functions representing the horizontal component of velocity below the partially immersed barriers and above the thick barrier. To approximate
the solutions, we employ the multi-term Galerkin technique using Chebyshev and
Gegenbauer polynomials as basis functions with appropriate weights. Our numerical results for reflection and transmission coefficients, presented graphically, are
validated through comparisons with existing literature [7].

2   Mathematical Formulation

In our study, we consider a finite-depth water body denoted as h, assuming it is
incompressible, inviscid, and homogeneous. We adopt the vertical y-axis downward
and represent the undisturbed free surface using the xz-plane. Our investigation
focuses on a configuration comprising a pair of partially immersed thin vertical
barriers and a thick bottom-standing barrier. These barriers partition the water body
into distinct regions with wetted barrier so that the wetted portion of the barrier
occupies the regions: $\Gamma_1 : -\infty < x < c_1, \ 0 < y < h$; $\Gamma_2 : c_1 < x < c_2, \ 0 < y < h$; $\Gamma_3 : c_2 < x < c_3, \ 0 < y < l$; $\Gamma_4 : c_3 < x < c_4, \ 0 < y < h$; $\Gamma_5 : c_4 < x < \infty, \ 0 < y < h$ (where $l = l_2 = l_3$, gap between the free surface and the top of
the thick barrier) as shown in Fig. 1.

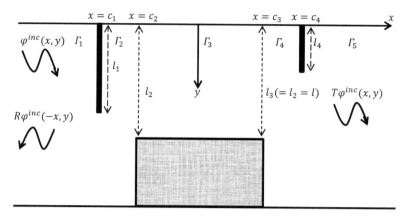

Fig. 1 Schematic diagram of the problem

Let a wave incident obliquely from the left side onto the barriers can be described as a train of surface waves propagating at an angle relative to the normal of the barriers is represented by $\text{Re}[\phi^{inc}(x, y)e^{-i\omega t + i\eta z}]$, where

$$\phi^{inc}(x, y) = \zeta_0(y)e^{i\mu(x-c_1)},$$

where $\zeta_0(y) = \frac{\cosh k_0(h-y)}{\cosh k_0 h}$ with $\eta = k_0 \sin\theta$, $\mu = k_0 \cos\theta$, k_0 is the unique real positive root of the dispersion relation $k \tanh kh = K$, $K = \frac{\omega^2}{g}$, ω is the circular frequency of the incident wave, g the acceleration due to gravity, and θ is the inclination angle in the regions Γ_1, Γ_2, Γ_4, and Γ_5. Due to Snell's law for refraction across discontinuities in the water depth, η satisfies

$$\eta = k_0 \sin\theta = \beta_0 \sin\theta' = \text{constant},$$

where β_0 is the wavenumber for the region Γ_3 and is the one and only real positive root of the transcendental relation $\beta \tanh \beta l = K$.

The velocity potential $\phi_j(x, y)$ satisfies the modified Helmholtz equation in each subregion,

$$(\nabla^2 - \eta^2)\phi_j = 0 \text{ in } \Gamma_j \ (j = 1, 2, 3, 4, 5) \tag{1}$$

along with the following conditions:

$$\frac{\partial \phi_j}{\partial y} + K\phi_j = 0, \text{ on } y = 0, \ x \in (-\infty, \infty) \tag{2}$$

$$\frac{\partial \phi_j}{\partial x} = 0 \text{ on } \begin{cases} x = c_j, \ y \in (0, l_j) \ (j = 1, 4) \\ x = c_j, \ y \in (l_j, h) \ (j = 2, 3) \end{cases} \tag{3}$$

$$\frac{\partial \phi_j}{\partial y} = 0 \text{ on } \begin{cases} y = h, \ x \in (-\infty, c_2) \cup (c_3, \infty) \ (j = 1, 2, 4, 5) \\ y = l, \ x \in (c_2, c_3) \ (j = 3) \end{cases} \quad (4)$$

$r^{1/2} \nabla \phi_j$ is bounded near submerged sharp edges of the thin barriers   (5a)

and

$r^{1/3} \nabla \phi_j$ is bounded near submerged sharp corners of the thick barrier,   (5b)

where r is the distance from submerged edges.

$$\phi(x, y) \sim \begin{cases} \phi^{inc}(x, y) + R\phi^{inc}(-x, y) \text{ as } x \to -\infty, \\ T\phi^{inc}(x, y) \text{ as } x \to \infty, \end{cases} \quad (6)$$

where R and T are reflection and transmission coefficients.

3   Solution Procedure

Using the methodology of Havelock's expansion to express the general solution for water wave potential $\phi_j(x, y)$ in each subdomain as

$$\phi_1(x, y) = \left\{ e^{i\mu(x-c_1)} + V_1 e^{-i\mu(x-c_1)} \right\} \frac{\cosh k_0(h - y)}{\cosh k_0 h} + \sum_{n=1}^{\infty} A_n e^{p_n(x-c_1)} \frac{\cos k_n(h - y)}{\cos k_n h}, \text{ in } \Gamma_1$$

$$\qquad\qquad (7a)$$

$$\phi_2(x, y) = \left\{ B_0 \cos p_0(x - c_1) + C_0 \sin p_0(x - c_2) \right\} \frac{\cosh k_0(h - y)}{\cosh k_0 h} +$$

$$\sum_{n=1}^{\infty} \left\{ B_n \cosh p_n(x - c_1) + C_n \sinh p_n(x - c_2) \right\} \frac{\cos k_n(h - y)}{\cos k_n h}, \text{ in } \Gamma_2$$

$$\qquad\qquad (7b)$$

$$\phi_3(x, y) = \left\{ D_0 \cos q_0(x - c_2) + E_0 \sin q_0(x - c_3) \right\} \frac{\cosh \beta_0(l - y)}{\cosh \beta_0 l} +$$

$$\sum_{n=1}^{\infty} \left\{ D_n \cosh q_n(x - c_2) + E_n \sinh q_n(x - c_3) \right\} \frac{\cos \beta_n(l - y)}{\cos \beta_n l}, \text{ in } \Gamma_3$$

$$\qquad\qquad (7c)$$

$$\phi_4(x, y) = \left\{ F_0 \cos p_0(x - c_3) + G_0 \sin p_0(x - c_4) \right\} \frac{\cosh k_0(h - y)}{\cosh k_0 h} +$$

$$\sum_{n=1}^{\infty} \left\{ F_n \cosh p_n(x - c_3) + G_n \sinh p_n(x - c_4) \right\} \frac{\cos k_n(h - y)}{\cos k_n h}, \text{ in } \Gamma_4$$

$$\qquad\qquad (7d)$$

$$\phi_5(x, y) = V_2 e^{i\mu(x-c_4)} \frac{\cosh k_0(h - y)}{\cosh k_0 h} + \sum_{n=1}^{\infty} H_n e^{-p_n(x-c_4)} \frac{\cos k_n(h - y)}{\cos k_n h}, \text{ in } \Gamma_5$$

(7e)

where $\pm i k_n$ $(n = 1, 2, ...)$ are purely imaginary roots of $k \tanh kh = K$; $\pm \beta_n$ $(n = 1, 2, ...)$ are of $\beta \tanh \beta l = K$, $p_n = \{(k_n)^2 + \eta^2\}^{1/2}$, $q_n = \{(\beta_n)^2 + \eta^2\}^{1/2}$, $p_0 = \{(k_0)^2 - \eta^2\}^{1/2}$ and $q_0 = \{(\beta_0)^2 - \eta^2\}^{1/2}$. Here $A_n, B_n, C_n, D_n, E_n, F_n, G_n, H_n$ $(n = 1, 2, ...)$ are unknown constants and V_1, V_2 are given by $V_1 = Re^{-2ic_1}$, $V_2 = Te^{i(c_4-c_1)}$.

Let us introduce

$$g_j(y) = \frac{\partial \phi_j}{\partial x}(c_j - 0, y) = \frac{\partial \phi_{j+1}}{\partial x}(c_j + 0, y), \ y \in (0, h) \ (j = 1, 2, 3, 4) \quad (8)$$

where $g_1(y) = 0$ for $y \in (0, l_1)$, $g_2(y) = 0$ for $y \in (l_2, h)$, $g_3(y) = 0$ for $y \in (l_3, h)$, and $g_4(y) = 0$ for $y \in (0, l_4)$ due to the condition (5).

Near the submerged edges of thin barriers, the horizontal components of velocity behave as

$$g_j(y) = O(|l_j - y|^{-\frac{1}{2}}) \text{ as } y \to l_j - 0 \ (j = 1, 4).$$

And near the submerged corners of thick barrier, the horizontal component of velocity behaves as

$$g_j(y) = O(|l_j - y|^{-\frac{1}{3}}) \text{ as } y \to l_j + 0 \ (j = 2, 3).$$

Let us choose the following constants:

$$A_1 = 1, \ A_2 = 0, \ A_3 = 0, \ A_4 = 0, \ B_1 = V_1, \ B_2 = E_0 \sin q_0(c_3 - c_2),$$
$$B_3 = G_0 \sin p_0(c_4 - c_3), \ B_4 = -V_2. \quad (9)$$

Now, we introduce the following step functions:

$$\mathcal{L}_j(y) = \begin{cases} 0, \ 0 < y < l_j, \\ 1, \ l_j < y < h; \end{cases} (j = 1, 4) \quad (10a)$$

$$\mathcal{L}_j(y) = \begin{cases} 1, \ 0 < y < l_j, \\ 0, \ l_j < y < h. \end{cases} (j = 2, 3). \quad (10b)$$

Utilizing Havelock's inversion formulae on $g_j(y)$ $(j = 1, 2, 3, 4)$ and employing the continuity of $\phi(x, y)$ at $x = c_1$, $y \in (l_1, h)$; $x = c_2$, $y \in (0, l_2)$; $x = c_3$, $y \in (0, l_3)$; $x = c_4$, $y \in (l_4, h)$, we have the following set of integral equations:

$$\sum_{i=1}^{4} \int_0^h \mathcal{L}_i(u)g_i(u)\mathcal{W}_{ji}(y, u)\mathcal{L}_j(y)dy = (A_j + B_j)\mathcal{Q}_{jj}, \ j = 1, 2, 3, 4; \quad (11)$$

where $Q_{jj} = \frac{\cosh k_0(h-y)}{\cosh k_0 h}$ for $j = 1, 3, 4$ and $Q_{jj} = \frac{\cosh \beta_0(l-y)}{\cosh \beta_0 l}$ for $j = 2$ and the kernels $\mathcal{W}_{ji}(y, u)$ $(j, i = 1, 2, 3, 4)$ are given by

$$\mathcal{W}_{11}(y, u) = \frac{4k_0 \cot p_0(c_2 - c_1) \cosh k_0(h - u) \cosh k_0(h - y)}{p_0(2k_0 h + \sinh 2k_0 h)} - \sum_{n=1}^{\infty} \frac{4k_n\{1 + \coth p_n(c_2 - c_1)\} \cos k_n(h - u) \cos k_n(h - y)}{p_n(2k_n h + \sin 2k_n h)} \tag{12a}$$

$$\mathcal{W}_{12}(y, u) = -\frac{4k_0 \csc p_0(c_2 - c_1) \cosh k_0(h - u) \cosh k_0(h - y)}{p_0(2k_0 h + \sinh 2k_0 h)} + \sum_{n=1}^{\infty} \frac{4k_n \operatorname{csch} p_n(c_2 - c_1) \cos k_n(h - u) \cos k_n(h - y)}{p_n(2k_n h + \sin 2k_n h)} \tag{12b}$$

$$\mathcal{W}_{22}(y, u) = \frac{8\beta_0 \csc 2q_0(c_3 - c_2) \cosh \beta_0(l - u) \cosh \beta_0(l - y)}{q_0(2\beta_0 l + \sinh 2\beta_0 l)} + \frac{4k_0 \cot p_0(c_2 - c_1) \cosh k_0(h - u) \cosh k_0(h - y)}{p_0(2k_0 h + \sinh 2k_0 h)} - \sum_{n=1}^{\infty} \frac{4\beta_n \coth q_n(c_3 - c_2) \cos \beta_n(l - u) \cos \beta_n(l - y)}{q_n(2\beta_n l + \sin 2\beta_n l)} - \sum_{n=1}^{\infty} \frac{4k_n \coth p_n(c_2 - c_1) \cos k_n(h - u) \cos k_n(h - y)}{p_n(2k_n h + \sin 2k_n h)} \tag{12c}$$

$$\mathcal{W}_{13}(y, u) = \mathcal{W}_{14}(y, u) = \mathcal{W}_{24}(y, u) = \mathcal{W}_{31}(y, u) = \mathcal{W}_{41}(y, u) = \mathcal{W}_{42}(y, u) = 0,$$
$$\mathcal{W}_{21}(y, u) = \mathcal{W}_{12}(y, u), \mathcal{W}_{23}(y, u) = \mathcal{W}_{32}(y, u), \mathcal{W}_{34}(y, u) = \mathcal{W}_{43}(y, u) \tag{12d}$$

$$\mathcal{W}_{23}(y, u) = -\frac{4\beta_0 \csc q_0(c_3 - c_2) \cosh \beta_0(l - u) \cosh \beta_0(l - y)}{q_0(2\beta_0 l + \sinh 2\beta_0 l)} + \sum_{n=1}^{\infty} \frac{4\beta_n \operatorname{csch} q_n(c_3 - c_2) \cos \beta_n(l - u) \cos \beta_n(l - y)}{q_n(2\beta_n l + \sin 2\beta_n l)} \tag{12e}$$

$$\mathcal{W}_{33}(y, u) = \frac{4\beta_0 \cot q_0(c_3 - c_2) \cosh \beta_0(l - u) \cosh \beta_0(l - y)}{q_0(2\beta_0 l + \sinh 2\beta_0 l)} + \frac{8k_0 \csc 2p_0(c_4 - c_3) \cosh k_0(h - u) \cosh k_0(h - y)}{p_0(2k_0 h + \sinh 2k_0 h)} - \sum_{n=1}^{\infty} \frac{4\beta_n \coth q_n(c_3 - c_2) \cos \beta_n(l - u) \cos \beta_n(l - y)}{q_n(2\beta_n l + \sin 2\beta_n l)} - \sum_{n=1}^{\infty} \frac{4k_n \coth p_n(c_4 - c_3) \cos k_n(h - u) \cos k_n(h - y)}{p_n(2k_n h + \sin 2k_n h)} \tag{12f}$$

$$W_{34}(y, u) = -\frac{4k_0 \csc p_0(c_4 - c_3) \cosh k_0(h - u) \cosh k_0(h - y)}{p_0(2k_0h + \sinh 2k_0h)} +$$

$$\sum_{n=1}^{\infty} \frac{4k_n \csch p_n(c_4 - c_3) \cos k_n(h - u) \cos k_n(h - y)}{p_n(2k_nh + \sin 2k_nh)} \tag{12g}$$

$$W_{44}(y, u) = \frac{4k_0 \cot p_0(c_4 - c_3) \cosh k_0(h - u) \cosh k_0(h - y)}{p_0(2k_0h + \sinh 2k_0h)} -$$

$$\sum_{n=1}^{\infty} \frac{4k_n \coth p_n(c_4 - c_3) \cos k_n(h - u) \cos k_n(h - y)}{p_n(2k_nh + \sin 2k_nh)} - \tag{12h}$$

$$\sum_{n=1}^{\infty} \frac{4k_n \cos k_n(h - u) \cos k_n(h - y)}{p_n(2k_nh + \sin 2k_nh)}$$

Also, we define the following transformation as

$$\mathbf{g}(u) = (\mathbf{A} + \mathbf{B})\mathbf{G}(u) \tag{13}$$

$\mathbf{g}(u) = \{g_j(u)\}_{4\times1}^T$, $\mathbf{A} = \{A_j\}_{4\times1}^T$, $\mathbf{B} = \{B_j\}_{4\times1}^T$, $\mathbf{G}(u) = \{G_{jl}(u)\}_{4\times4}$.

The set of integral equations in (11) can be represented in a single equation with the help of (9). Then applying the step functions as given by (10), we transform the ranges of integration into $(0, h)$. After that the conversion (13) helps to modify the integral equation in the following matrix:

$$\mathbf{L}(y)\mathbf{Q}(y) = \int_0^h \mathbf{L}(u)\mathbf{W}(y, u)\mathbf{L}(u)\mathbf{G}(u)du, \quad 0 < y < h, \tag{14}$$

where $\mathbf{G}(y) = \text{diag}\big(\mathcal{G}_j(y)\big)_{4\times4}$, $\mathbf{Q}(y) = \text{diag}\big(\mathcal{Q}_{jj}(y)\big)_{4\times4}$, $\mathbf{W}(y, u) = \big(\mathcal{W}_{ji}(y, u)\big)_{4\times4}$.

Using Havelock's inversion formulae, the unknown constants V_1, E_0, G_0, and V_2 are represented in terms of the horizontal component of velocities. Further using (9), (10), and (13), we get the following matrix equation:

$$\mathbf{B} = \mathbf{XA}, \tag{15}$$

where $\mathbf{X} = \big[\mathbf{MU} + \mathbf{Z}\big]^{-1}\big[\mathbf{MU} - \mathbf{Z}\big]$, $\mathbf{M} = \text{diag}\big(\frac{2k_0h + \sinh 2k_0h}{4\cosh^2 k_0h}, \frac{2\beta_0l + \sinh 2\beta_0l}{4\cosh^2 \beta_0l},$
$\frac{2k_0h + \sinh 2k_0h}{4\cosh^2 k_0h}, \frac{2k_0h + \sinh 2k_0h}{4\cosh^2 k_0h}\big)$, $\mathbf{U} = \text{diag}\big(\frac{i\mu}{k_0}, \frac{-q_0 \cot q_0(c_3 - c_2)}{\beta_0}, \frac{-p_0 \cot p_0(c_4 - c_3)}{k_0}, \frac{i\mu}{k_0}\big)$ and $\mathbf{Z}$ is of the form

$$\mathbf{Z} = \int_0^h \mathbf{L}(u)\mathbf{Q}(u)\mathbf{G}(u)du. \tag{16}$$

If we get $\mathbf{X}$, then the elements V_1 and V_2 of B are obtained by solving (15) for $\mathbf{B}$. So that, reflection and transmission coefficients R and T, related to V_1 and V_2, respectively, are determined.

4   Multi-term Galerkin Technique

Evaluating the value of $\mathbf{Z}$ we can find $\mathbf{X}$. Thus for that, $(N+1)$-term Galerkin approximation of $G_{jl}(u)$ ($\mathbf{G}(u) = \{G_{jl}(u)\}_{4 \times 4}$) is considered as

$$G_{jl}(u) \simeq \sum_{n=0}^{N} c_n^{(jl)} \chi_n^{(j)}(u), \quad j, l = 1, 2, 3, 4, \tag{17}$$

where $c_n^{(jl)}$ and $\chi_n^{(j)}(u)$ are unknown constants and the basis functions, respectively. The basis functions $\chi_n^{(j)}(y)$'s are taken as

$$\chi_n^{(j)}(y) = \begin{cases} -\frac{d}{dy}\left[e^{-Ky} \int_y^{l_j} e^{Kt} \hat{\chi}_n^{(j)}(t) dt \right], & y \in (l_j, h) \ (j = 2, 3), \\ \chi_n^{(j)}(y), & y \in (0, l_j) \ (j = 1, 4), \end{cases} \tag{18}$$

with

$$\hat{\chi}_n^{(j)}(t) = \begin{cases} \dfrac{2(-1)^n}{\pi\left\{(h-l_j)^2-(h-t)^2\right\}^{\frac{1}{2}}} T_{2n}\left(\dfrac{h-t}{h-l_j}\right), & t \in (l_j, h) \ (j = 1, 4), \\[6mm] \dfrac{2^{\frac{7}{6}}\Gamma(\frac{1}{6})(2n)!}{\pi\Gamma(2n+\frac{1}{3}) l_j^{\frac{1}{3}} (l_j^2-t^2)^{\frac{1}{3}}} C_{2n}^{\frac{1}{6}}\left(\dfrac{t}{l_j}\right), & t \in (0, l_j) \ (j = 2, 3), \end{cases} \tag{19}$$

where $T_{2n}(x)$ is the Chebychev polynomial of first kind of order $2n$ and $C_{2n}^{\frac{1}{6}}(x)$ the ultra-spherical Gegenbauer polynomial of order $\frac{1}{6}$.

Now substituting (17) into (14), we have four integral equations. On multiplication of jth integral equation by $\chi_m^{(j)}(y)$ ($j = 1, 2, 3, 4$) and integration over $(0, h)$ successively, the following system of equations is obtained as

$$\sum_{n=0}^{N} \sum_{j=1}^{4} c_n^{(jl)} \mathcal{F}_{ij}^{(mn)} = \delta_{il} \mathcal{U}_i^{(m)}, \quad i = 1, 2, 3, 4; \ m = 0, 1, ..., N, \tag{20}$$

where, with the help of $\mathcal{W}_j(y, u)$ ($j = 1(1)10$) given in equation (10), $\mathcal{F}_{ij}^{(mn)}, \mathcal{U}_i^{(m)}$ ($i, j = 1, 2, 3, 4; m, n = 0, 1, ..., N$) are of written in the following form as

$$\mathcal{F}_{11}^{(mn)} = \frac{4k_0 \cot p_0 (c_2 - c_1)}{p_0(2k_0 h + \sinh 2k_0 h)}(-1)^{m+n} \mathcal{I}_{2m}\{k_0(h-l_1)\}\mathcal{I}_{2n}\{k_0(h-l_1)\}-$$
$$\sum_{r=1}^{\infty} \frac{4k_r\{1 + \coth p_r(c_2 - c_1)\}}{p_r(2k_r h + \sin 2k_r h)} \mathcal{J}_{2m}\{k_r(h-l_1)\}\mathcal{J}_{2n}\{k_r(h-l_1)\} \tag{21a}$$

$$
\begin{aligned}
\mathcal{F}_{12}^{(mn)} ={}& \frac{-8k_0 \csc p_0 (c_2 - c_1) \cosh k_0 l}{p_0 (2k_0 h + \sinh 2k_0 h)} (-1)^m \mathcal{I}_{2m}\{k_0(h - l_1)\} \frac{\mathcal{I}_{2n+\frac{1}{6}}(k_0 l_2)}{(k_0 l_2)^{\frac{1}{6}}} \\
&+ \sum_{r=1}^{\infty} \frac{8k_r \operatorname{csch} p_r (c_2 - c_1)}{p_r (2k_r h + \sin 2k_r h)} (-1)^n \mathcal{J}_{2m}\{k_r(h - l_1)\} \frac{\mathcal{J}_{2n+\frac{1}{6}}(k_r l_2)}{(k_r l_2)^{\frac{1}{6}}}
\end{aligned}
\tag{21b}
$$

$$
\mathcal{F}_{13}^{(mn)} = \mathcal{F}_{14}^{(mn)} = 0, \quad \mathcal{F}_{21}^{(mn)} = \mathcal{F}_{12}^{(nm)}
\tag{21c}
$$

$$
\begin{aligned}
\mathcal{F}_{22}^{(mn)} ={}& \frac{32\beta_0 \csc 2q_0 (c_3 - c_2) \cosh^2 \beta_0 l}{q_0 (2\beta_0 l + \sinh 2\beta_0 l)} \frac{\mathcal{I}_{2m+\frac{1}{6}}(\beta_0 l_2) \mathcal{I}_{2n+\frac{1}{6}}(\beta_0 l_2)}{(\beta_0 l_2)^{\frac{1}{3}}} + \\
& \frac{16k_0 \cot p_0 (c_2 - c_1) \cosh^2 k_0 l}{p_0 (2k_0 h + \sinh 2k_0 h)} \frac{\mathcal{I}_{2m+\frac{1}{6}}(k_0 l_2) \mathcal{I}_{2n+\frac{1}{6}}(k_0 l_2)}{(k_0 l_2)^{\frac{1}{3}}} - \\
& \sum_{r=1}^{\infty} \frac{16\beta_r \coth q_r (c_3 - c_2)}{q_r (2\beta_r l + \sin 2\beta_r l)} (-1)^{m+n} \frac{\mathcal{J}_{2m+\frac{1}{6}}(\beta_r l_2) \mathcal{J}_{2n+\frac{1}{6}}(\beta_r l_2)}{(\beta_r l_2)^{\frac{1}{3}}} - \\
& \sum_{r=1}^{\infty} \frac{16k_r \coth p_r (c_2 - c_1)}{p_r (2k_r h + \sin 2k_r h)} (-1)^{m+n} \frac{\mathcal{J}_{2m+\frac{1}{6}}(k_r l_2) \mathcal{J}_{2n+\frac{1}{6}}(k_r l_2)}{(k_r l_2)^{\frac{1}{3}}}
\end{aligned}
\tag{21d}
$$

$$
\begin{aligned}
\mathcal{F}_{23}^{(mn)} ={}& \frac{-16\beta_0 \csc q_0 (c_3 - c_2) \cosh^2 \beta_0 l}{q_0 (2\beta_0 l + \sinh 2\beta_0 l)} \frac{\mathcal{I}_{2m+\frac{1}{6}}(\beta_0 l_2) \mathcal{I}_{2n+\frac{1}{6}}(\beta_0 l_3)}{(\beta_0 l_2)^{\frac{1}{6}} (\beta_0 l_3)^{\frac{1}{6}}} \\
&+ \sum_{r=1}^{\infty} \frac{16\beta_r \operatorname{csch} q_r (c_3 - c_2)}{q_r (2\beta_r l + \sin 2\beta_r l)} (-1)^{m+n} \frac{\mathcal{J}_{2n+\frac{1}{6}}(\beta_r l_2) \mathcal{J}_{2n+\frac{1}{6}}(\beta_r l_3)}{(\beta_r l_2)^{\frac{1}{6}} (\beta_r l_3)^{\frac{1}{6}}}
\end{aligned}
\tag{21e}
$$

$$
\mathcal{F}_{24}^{(mn)} = \mathcal{F}_{31}^{(mn)} = 0, \quad \mathcal{F}_{32}^{(mn)} = \mathcal{F}_{23}^{(nm)}
\tag{21f}
$$

$$
\begin{aligned}
\mathcal{F}_{33}^{(mn)} ={}& \frac{32k_0 \csc 2p_0 (c_4 - c_3) \cosh^2 k_0 h}{p_0 (2k_0 h + \sinh 2k_0 h)} \frac{\mathcal{I}_{2m+\frac{1}{6}}(k_0 l_3) \mathcal{I}_{2n+\frac{1}{6}}(k_0 l_3)}{(k_0 l_3)^{\frac{1}{3}}} + \\
& \frac{16\beta_0 \cot q_0 (c_3 - c_2) \cosh^2 \beta_0 l}{q_0 (2\beta_0 l + \sinh 2\beta_0 l)} \frac{\mathcal{I}_{2m+\frac{1}{6}}(\beta_0 l_3) \mathcal{I}_{2n+\frac{1}{6}}(\beta_0 l_3)}{(\beta_0 l_3)^{\frac{1}{3}}} - \\
& \sum_{r=1}^{\infty} \frac{16\beta_r \coth q_r (c_3 - c_2)}{q_r (2\beta_r l + \sin 2\beta_r l)} (-1)^{m+n} \frac{\mathcal{J}_{2m+\frac{1}{6}}(\beta_r l_3) \mathcal{J}_{2n+\frac{1}{6}}(\beta_r l_3)}{(\beta_r l_3)^{\frac{1}{3}}} - \\
& \sum_{r=1}^{\infty} \frac{16k_r \coth p_r (c_4 - c_3)}{p_r (2k_r h + \sin 2k_r h)} (-1)^{m+n} \frac{\mathcal{J}_{2m+\frac{1}{6}}(k_r l_3) \mathcal{J}_{2n+\frac{1}{6}}(k_r l_3)}{(k_r l_3)^{\frac{1}{3}}}
\end{aligned}
\tag{21g}
$$

$$
\begin{aligned}
\mathcal{F}_{34}^{(mn)} ={}& \frac{-8k_0 \csc p_0 (c_4 - c_3) \cosh k_0 l}{p_0 (2k_0 h + \sinh 2k_0 h)} \mathcal{I}_{2n}\{k_0(h - l_4)\} \frac{\mathcal{I}_{2m+\frac{1}{6}}(k_0 l_3)}{(k_0 l_3)^{\frac{1}{6}}} + \\
& \sum_{r=1}^{\infty} \frac{8k_r \operatorname{csch} p_r (c_4 - c_3)}{p_r (2k_r h + \sin 2k_r h)} (-1)^m \mathcal{J}_{2n}\{k_r(h - l_4)\} \frac{\mathcal{J}_{2m+\frac{1}{6}}(k_r l_3)}{(k_r l_3)^{\frac{1}{6}}}
\end{aligned}
\tag{21h}
$$

$$\mathcal{F}_{41}^{(mn)} = \mathcal{F}_{42}^{(mn)} = 0, \ \mathcal{F}_{34}^{(mn)} = \mathcal{F}_{43}^{(nm)} \tag{21i}$$

$$\mathcal{F}_{44}^{(mn)} = \frac{4k_0 \cot p_0 (c_4 - c_3)}{p_0 (2k_0 h + \sinh 2k_0 h)} (-1)^{m+n} \mathcal{I}_{2m}\{k_0 (h - l_4)\} \mathcal{I}_{2n}\{k_0 (h - l_4)\} -$$
$$\sum_{r=1}^{\infty} \frac{4k_r \{1 + \coth p_r (c_4 - c_3)\}}{p_r (2k_r h + \sin 2k_r h)} \mathcal{J}_{2m}\{k_r (h - l_4)\} \mathcal{J}_{2n}\{k_r (h - l_4)\} \tag{21j}$$

$$\mathcal{U}_j^{(m)} = (-1)^m \frac{\mathcal{I}_{2m}\{k_0 (h - l_j)\}}{\cosh k_0 h} \ (j = 1, 4) \tag{21k}$$

$$\mathcal{U}_j^{(m)} = \frac{2\mathcal{I}_{2m+\frac{1}{6}}(\beta_0 l_j)}{(\beta_0 l_j)^{\frac{1}{6}}} \ (j = 2, 3) \tag{21l}$$

where $\mathcal{J}_n$, $\mathcal{I}_n$ are the Bessel function and modified Bessel function of order n, respectively.

After getting of $\mathcal{F}_{ij}^{(mn)}, \mathcal{U}_i^{(m)} \ (i, j = 1, 2, 3, 4)$, Eq. (20) gives $c_n^{(jl)}$ and then assuming $\mathbf{Z} = \{Z_{jl}\}_{4 \times 4}$ so that Z_{jl}'s $(j, l = 1, 2, 3, 4)$ are obtained from (16) as

$$Z_{jl} \simeq \sum_{n=0}^{N} c_n^{(jl)} \mathcal{U}_j^{(n)}, \ j, l = 1, 2, 3, 4. \tag{22}$$

5 Numerical Results

In this section, a Wolfram Mathematica program is developed to scrutinize the effects of various wave and structural parameters on wave scattering. We aim to analyze the impact of factors such as angle of incidence, submerged gap of a thick bottom-standing barrier, lengths of thin barriers, and width of a thick barrier on $|R|$ and $|T|$. Through numerical results and analysis, we assess the effectiveness of using a combination of thick and thin barriers as a breakwater.

To ratify our current findings with the results obtained by Mandal and Kanoria [7], we set $\frac{l_1}{h} = 0.0001 = \frac{l_4}{h}$ to adequately abolish the impact of pair of thin immersed barriers. Therefore, our Fig. 2 closely matches with the graph of Fig. 6 of Mandal and Kanoria [7] which focused on a thick bottom-standing barrier. To achieve this resemblance, we selected specific parameters: $\frac{l_2}{h} = 0.5 (= \frac{l_3}{h})$, $\frac{c_1}{h} = 0.0001$, $\frac{c_2}{h} = 0.00011$, $\frac{c_3}{h} = 0.000111$, $\frac{c_4}{h} = 0.2$, and $\theta = 45°$. This comparison serves as evidence that the numerical method employed in our current study produces highly accurate results.

Figure 3 illustrates the curves depicting the absolute estimations of $|R|$ and $|T|$ for various angles of incidence of propagating waves. Figure 3a, b shows that both $|R|$ and $|T|$ become more oscillatory when the angle of incidence decreases. Also,

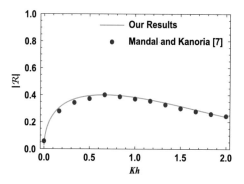

Fig. 2 $\frac{l_2}{h} = 0.5(= \frac{l_3}{h})$, $\frac{c_1}{h} = 0.0001$, $\frac{c_2}{h} = 0.00011$, $\frac{c_3}{h} = 0.000111$, $\frac{c_4}{h} = 0.2$ and $\theta = 45°$

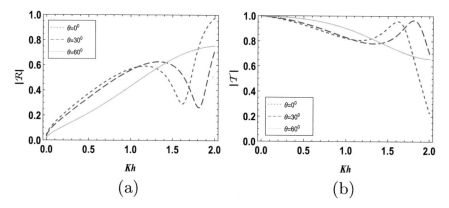

Fig. 3 $|R|$ **a** and $|T|$ **b** plotted against Kh for different angles of incidence with $\frac{l_1}{h} = 0.4$, $\frac{l_2}{h} = 0.6(= \frac{l_3}{h})$, $\frac{l_4}{h} = 0.3$, $\frac{c_1}{h} = 0.25$, $\frac{c_2}{h} = 0.5$, $\frac{c_3}{h} = 0.75$, $\frac{c_4}{h} = 1.0$

it is observed that maxima of $|R|$ and minima of $|T|$ are shifted toward the right as the angle of incidence increases.

Figure 4 showcases multiple graphs representing the $|R|$ and $|T|$ for different widths of the thick barrier. A contrasting graphical representation is visualized for $|R|$ and $|T|$ in Fig. 4a, b respectively. It is inferred that complete reflection and zero transmission occur for some discrete frequencies. Furthermore, when the width of a thick barrier is increased, $|R|$ and $|T|$ exhibit reduced oscillations. This can be attributed to the fact that a wider barrier allows for a greater range of energy states to be accommodated, resulting in a more gradual transition for the incident energy across the barrier.

In Fig. 5, the influence of the length of thin barriers is manifested under the consideration of three different lengths of barriers. Figure 5a reveals a transition in the reflection behavior as the barrier length increases, with an increase in reflection for $Kh < 1.4$ and a decrease in reflection for $Kh > 1.4$. But in Fig. 5a, b, the amount of transmission has a contrasting nature to this fact that occurs when $Kh > 1.4$.

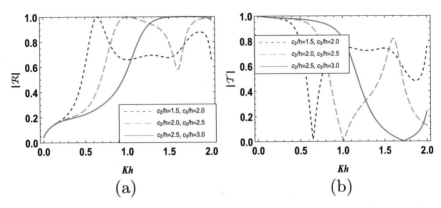

Fig. 4 $|R|$ **a** and $|T|$ **b** plotted against Kh for different widths of thick barrier with $\frac{l_1}{h} = 0.35$, $\frac{l_2}{h} = 0.7(= \frac{l_3}{h})$, $\frac{l_4}{h} = 0.45$, $\frac{c_1}{h} = 1.0$, $\frac{c_4}{h} = 4.0$, $\theta = 30°$

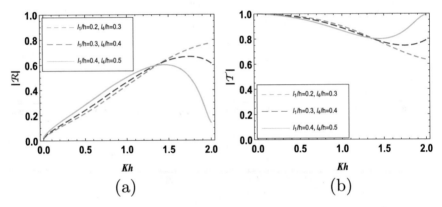

Fig. 5 $|R|$ **a** and $|T|$ **b** plotted against Kh for different lengths of thin barrier with $\frac{l_2}{h} = 0.65(= \frac{l_3}{h})$, $\frac{c_1}{h} = 0.4$, $\frac{c_2}{h} = 0.8$, $\frac{c_3}{h} = 1.2$, $\frac{c_4}{h} = 1.6$, $\theta = 60°$

Figure 6a demonstrates that with the decrease of the submerged gap of the thick barrier, there is an increase in the peak value of $|R|$. This phenomenon is attributed due to the large size of the thick barrier that reflected more wave energy. Meanwhile, a contrasting graphical representation is visualized for $|T|$ in Fig. 6b.

In Fig. 7, $|R|$ and $|T|$ are plotted as a function of wavenumber Kh for three different cases viz. (i) lengths of both the thin barriers less than submerged gap of thick barrier ($\frac{l_1}{h} = 0.33(= \frac{l_4}{h})$, $\frac{l_2}{h} = 0.55(= \frac{l_3}{h})$), (ii) length of one the thin barrier greater than and other less than submerged gap of thick barrier ($\frac{l_1}{h} = 0.77$, $\frac{l_2}{h} = 0.55(= \frac{l_3}{h})$, $\frac{l_4}{h} = 0.33$), and (iii) lengths of both the thin barriers greater than submerged gap of thick barrier ($\frac{l_1}{h} = 0.77(= \frac{l_4}{h})$, $\frac{l_2}{h} = 0.55(= \frac{l_3}{h})$). It is observed from Fig. 7a, b that $|R|$ and $|T|$ become more oscillatory for case (iii) than the other two cases.

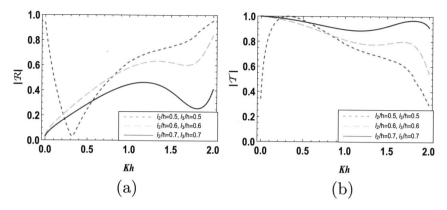

Fig. 6 $|R|$ **a** and $|T|$ **b** plotted against Kh for different submerged gaps of thick barrier with $\frac{l_1}{h} = 0.3(= \frac{l_4}{h})$, $\frac{c_1}{h} = 0.5$, $\frac{c_2}{h} = 1.0$, $\frac{c_3}{h} = 1.5$, $\frac{c_4}{h} = 2.0$, $\theta = 45°$

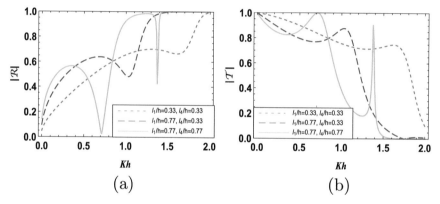

Fig. 7 $|R|$ **a** and $|T|$ **b** plotted against Kh for different lengths of thin barriers with $\frac{l_2}{h} = 0.55(= \frac{l_3}{h})$, $\frac{c_1}{h} = 0.3$, $\frac{c_2}{h} = 0.6$, $\frac{c_3}{h} = 0.9$, $\frac{c_4}{h} = 1.2$, $\theta = 0°$

6 Conclusions

This paper herein presents the $|R|$ and $|T|$ for a pair of thin vertical partially immersed barriers along with a thick bottom-standing barrier under the action of oblique incident waves. On the basis of multi-term Galerkin approximation technique, a set of integral equations with two types of singularities viz. half-singularity and one-third singularity is solved by choosing appropriate basis functions with suitable weights. Numerical and graphical results show the $|R|$ and $|T|$ for different parameters in a breakwater system. The angle of incidence, width of thick barrier, and length of thin barriers are crucial for an efficient breakwater design. The size of the thick barrier significantly affects the construction. The peak of $|R|$ depends on the increase of submerged gaps of the thick barrier. $|R|$ and $|T|$ behave less oscillatory nature for the case where lengths of thin barriers are less than submerged gap of thick barrier.

The study successfully handles square-root and one-third singularities in integral equations. It is applicable for investigating various barrier configurations with thick and thin barriers for wave scattering in shallow and deep water.

References

1. Tuck EO (1971) Transmission of water waves through small apertures. J Fluid Mech 49:481–491
2. Packham BA, Williams WE (1972) A note on the transmission of water waves through small apertures. J Inst Maths Applics 10:176–184
3. Mandal BN (1987) A note on the diffraction of water waves by a vertical wall with a narrow gap. Arch Mech 39:269–273
4. Mei CC, Black JL (1969) Scattering of surface waves by rectangular obstacles in waters of finite depth. J Fluid Mech 38:499–511
5. Guiney DC, Noye BJ, Tuck EO (1972) Transmission of water waves through small apertures. J Fluid Mech 55:149–161
6. Kanoria M, Dolai DP, Mandal BN (1999) Water wave scattering by thick vertical barriers. J Eng Math 35:361–384
7. Mandal BN, Kanoria M (2000) Oblique wave- scattering by thick horizontal barriers. J Offshore Mech Arctic Eng 122:100–108
8. Paul S, De S (2021) Interaction of flexural gravity wave in ice cover with a pair of bottom-mounted rectangular barriers. Ocean Eng 220:108449
9. Sasmal A, De S (2022) Analysis of oblique wave diffraction by rectangular thick barrier in the presence of surface tension. Indian J Phys 96:2051–2063
10. Ursell F (1947) The effect of a fixed vertical barrier on surface waves in deep water. Proc Camb Philos Soc 43:374–382
11. Levine H, Rodemich E (1958) Scattering of surface waves on an ideal fluid. Math Stat Lab Tech Rep 78:1–64
12. Das P, Dolai DP, Mandal BN (1997) Oblique wave diffraction by parallel thin vertical barriers with gaps. J Waterway Port Coastal Ocean Eng 123:163–171

Secure and Privacy-Compliant Data Sharing: An Essential Framework for Healthcare Organizations

Redwan Walid, Karuna Pande Joshi, and Lavanya Elluri

Abstract Data integration from multiple sources can improve decision-making and predict epidemiological trends. While there are many benefits to data integration, there are also privacy concerns, especially in healthcare. The Health Insurance Portability and Accountability Act (HIPAA) is one of the essential regulations in healthcare, and it sets strict standards for the privacy and security of patient data. Often, data integration can be complex because different rules apply to different companies. Many existing data integration technologies are domain-specific and theoretical, while others rigorously adhere to unified data integration. Moreover, the integration systems do not have semantic access control, which causes privacy breaches. We propose a framework using a knowledge graph for sharing and integrating data across healthcare providers by protecting data privacy. We use an ontology to provide Attribute-Based Access Control (ABAC) for preventing excess or unwanted access based on the user attributes or central organization rules. The data is shared by removing sensitive attributes and anonymizing the rest using k-anonymity to strike a balance between data utility and secret information. A metadata layer describes the schema mapping to integrate data from multiple sources. Our framework is a promising approach to data integration in healthcare, and it addresses some of the critical challenges of data integration in this domain.

Keywords Privacy · Security · Electronic Health Record (EHR) · Knowledge graph (ontology) · Attribute-Based Access Control (ABAC) · k-anonymity · Metadata

R. Walid (✉) · K. P. Joshi
Department of Information Systems, University of Maryland, Baltimore County, MD, USA
e-mail: rwalid1@umbc.edu

K. P. Joshi
e-mail: kjoshi1@umbc.edu

L. Elluri
Subhani Department of Computer Information Systems, Texas A&M University-Central Texas, Killeen, TX, USA
e-mail: elluri@tamuct.edu

© The Author(s), under exclusive license to Springer Nature Singapore Pte Ltd. 2024
D. Giri et al. (eds.), *Proceedings of the Tenth International Conference on Mathematics and Computing*, Lecture Notes in Networks and Systems 964,
https://doi.org/10.1007/978-981-97-2066-8_2

15

1 Introduction

The value of data in scientific experiments within healthcare is immeasurable, especially as the amount of data being generated continues to multiply. Different entities often create this data independently, and integrating it is crucial to transforming it into useful information. The integration allows organizations to combine data from various sources, providing consumers with real-time insight into business outcomes, increasing productivity, improving decision-making, and enabling future predictions. While data reside in separate locations, integration is the first step toward converting data into valuable and relevant information. According to [1], the data integration system aims to provide a uniform query interface to many data sources. It can help the user from having to locate each source, interact with each source in isolation, and then combine the results.

Although data integration has numerous benefits, it also presents several challenges related to data privacy. Data privacy refers to an individual's right to control data exchange with third parties within a computer network. Data integration-related privacy concerns can be significant. In the healthcare industry, for instance, the Health Insurance Portability and Accountability Act (HIPAA) is a crucial privacy regulation that safeguards patient data in the United States [2]. Other domains have similar privacy regulations, such as the General Data Protection Regulation (GDPR), which restricts what companies can do with customer information [3]. Data providers establish security protocols that define specific criteria for acquiring, analyzing, and sharing personal and confidential data during integration. If integration systems fail to implement these security protocols properly, data leakage risks and other attacks may occur, resulting in substantial legal and financial consequences. It is, therefore, essential to implement security measures that determine who can access data for specific purposes and consider the implications of exposing such data to users.

Integrating data from various databases can be complicated due to different policies that apply across multiple organizations. The process typically involves identifying data sources, creating a mediated schema, mapping data, formulating and executing queries, and displaying the results. Nevertheless, integration systems often fail to consider users' access levels based on their attributes or roles. For example, a senior doctor may have read-and-write access to all fields in a patient's Electronic Health Record (EHR), while a junior doctor may only have read access to specific fields. While data integration technologies are available, they are often domain-specific and theoretical, and some only strictly adhere to unified data integration. Currently, no standard model in the literature considers data privacy during integration. Although many scholars have been working on data integration, only a few have addressed the challenge of data privacy.

1.1 Motivation

There have been various applications in the physical world that involve data integration while following strict privacy constraints. One such application is described below.

Analysis of disease occurrence, frequency, and risk factors is essential to identifying and managing them. These analyses have an enormous effect on policy decisions. An apparent precondition for such analyses is the availability of the necessary data. The data must then be gathered and integrated by various healthcare facilities while sanitizing information vulnerable to privacy. Privacy concerns are a significant barrier to standardizing such activities. Infringement of privacy can inflict severe physical and mental distress on entities. The privacy-conserving integration and collaboration of experimental data in the health sciences have become critical to encouraging scientific exploration.

Considering the above application scenario, this work proposes a framework for sharing and integrating data across healthcare providers by protecting data privacy.

1.2 Our Approach

The proposed framework ensures safe sharing and data integration between various sources using different components, as listed below. An organization can exchange user data with other organizations, guaranteeing no violation of privacy.

1. The system uses HIPAA-compliant knowledge graph that uses W3C Web Ontology Language (OWL) [4] to provide Attribute-Based Access Control (ABAC) [5]. The knowledge graph helps control data leakage and provides limited user access based on user attributes and queries.
2. The data sources remove the sensitive attributes and then anonymize the data using k-anonymity [6]. K-anonymity is a technique developed to mitigate the possibility of anonymized data being re-identified when connecting to other datasets. The main objective is to anonymize the data shared by the data sources to protect privacy.
3. The system built a metadata layer on top of the data sources. The layer contains information about canonical names and names used in the local databases. The layer is used for data integration from the databases for semantic mapping. For any query, the data is derived by running subqueries from the origins and compiling them into a unified, coherent data set.

The remainder of the paper is structured as follows—we discuss related work in Section 2, System Overview in Section 3, Dataset Description in Section 4, Implementation in Section 5, and conclusion in Section 6.

2 Related Work

There have been several works done to integrate healthcare data. In [7], Bahga et al. developed a framework to integrate data from distributed and heterogeneous sources into a standard terminology in the cloud environment. Their approach does not consider privacy issues. In [8], Clifton et al. proposed a privacy framework for data integration. They identified a few potential research directions and challenges that must be considered to perform privacy-preserving data integration. In [9], the authors proposed a framework in a development state, and they also identified a few research directions. Lu et al. in [10] proposed privacy solutions for smart healthcare systems. They examined identification, access control, and detection research practices and concentrated on EHR's critical roles and features. In [11], Yau et al. presented a repository protecting privacy to integrate data from multiple data storage providers. The repository receives only the necessary amount of information from data sources depending on user requests for integration and rejects other claims. The limitation of their work is that they solely focused on matching operations.

Database researchers have also been scaling up development efforts to make privacy a core concern; for example, the Hippocratic project databases [12] and Platform for Privacy Preferences (P3P) [13]. When data is shared across various entities and transformed and integrated with other data sources, none of these initiatives answer privacy issues. In [14], the authors have suggested using a trusted third party for privacy-preserving data integration. However, they have not specified how trustworthy is the third party or if there is an evaluation metric. Organizations typically don't want to share data without such considerations. A variety of research efforts have been made on the development of data mining algorithms that protect privacy. One method adopts a distributed framework [15–17]; the other applies random noise to the data while retaining the distribution's underlying properties [18, 19]. These strategies presume data were integrated before their operations. More often, sources will not be amenable to give their data provided; data protection is maintained [8].

Several other works have been done on solving data integration challenges, and most are domain-specific [20–23]. Not many studies discussed policy issues that could impact data integration across various organizations. In [20], David addressed the data integration challenges in the omics platform, which is very challenging because the data is highly diversified, particularly in a drug discovery business that intrinsically pulls on many divergent data types. In [24], Alshawi et al. proposed a framework for assessing the quality and integration of patient data in the healthcare industry for Customer Relationship Management applications. In [25], Rigby et al. proposed a prototype knowledge broker that gathers and incorporates patient data from independent healthcare organizations using a software service model, providing a few solutions to enterprise-based file systems. Several data privacy-related works include [6, 26, 27]. The most popular methods are k-anonymity [6] and differential privacy [27]. Most of them consider privacy issues while sharing the data. Still, none considers a standard model or framework that concerns privacy issues with the data while integrating data from disparate sources, which might be very significant in a few domains like the healthcare industry, human resource recruiting companies, and financial institutions.

3 System Overview

Our proposed system has multiple layers depicted in the high-level system diagram in Fig. 1. Assume three hospitals have planned to share their local databases with a research organization. The data sources are identified as one, two, and three. An example of how the framework will work is: the user queries the framework, the user gets authenticated, and attributes get checked with a knowledge graph, the data

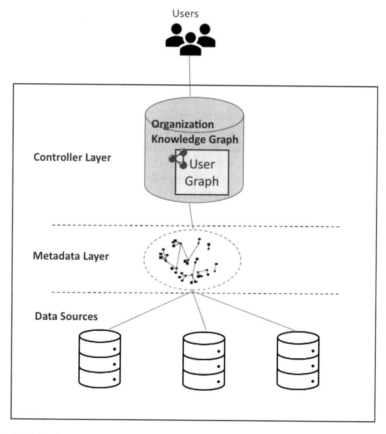

Fig. 1 High-level system diagram

sources remove the identifier and confidential attributes, the data sources anonymize the remaining data, subqueries get executed on the data sources, and the results are combined and displayed.

3.1 System Description

The system comprises multiple components within each module, as shown in the system architecture in Fig. 2. We describe each module in detail in the following few subsections.

3.1.1 Access Control Module

Attribute-Based Access Control is the primary function of the module. ABAC offers flexible, context-conscious, and cost-intelligent resource access management [28]. It allows access control strategies to require unique attributes from several separate systems to address permission and achieve successful regulatory enforcement, enabling organizations to be agile in their execution, depending on current resources. It follows Binary logic, with regulations including "IF, THEN" statements on who makes the order, the property, and the behavior.

When a user requests access, their username and password get checked with the database, and their attributes are verified. They will be denied access to the data if they do not fulfill them. Organization-defined policies are written in the Policy Controller in terms of user attributes. For example, a Senior Doctor having attributes *Certification:* CMA (Certified Medical Assistant), *Authority:* AAMA (American Association Of Medical Assistants), *Medical Institution:* GBMC (Greater Baltimore Medical Center), *Specialization:* AI (Allergy And Immunology), *Professional Designation:* Doctor will have write access to all the fields of integrated data. Likewise, a Junior Doctor will have partial access to the integrated data and may only read the data.

We use a HIPAA-compliant knowledge graph that uses W3C Web Ontology Language (OWL) [4] to control integrated data access. The knowledge graph was built

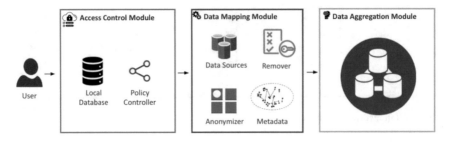

Fig. 2 System architecture

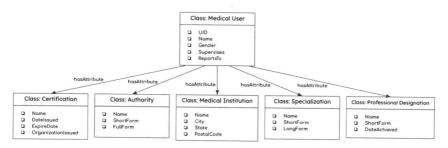

Fig. 3 Knowledge graph used in the system

by Joshi et al. [29], complying with the HIPAA standard [52], and it was transformed accordingly to capture the experimental scenario. Figure 3 shows the picture of the knowledge graph we used in our system. The knowledge graph uses the user's attribute to control access according to the policy defined by the organization. The knowledge graph may also be used to incorporate organization-specific access regulations. It can also control access to the fields of integrated data. For example, an EHR has several fields like Billing Information, Diagnosis, Allergies, etc., and the knowledge graph can be used to control access to each field of an EHR. Likewise, in our application scenario, we use the knowledge graph to control access to the fields of the integrated data.

3.1.2   Data Mapping Module

Once a requestor passes the access control module, the data sources get identified. The data sources remove the sensitive attribute with the help of the remover. The anonymizer anonymizes the remaining data to be shared to protect privacy. Anonymization is the way of altering data until it is released for data processing. Therefore, de-identification is not achievable and would result in k identifiable information if an effort is made to de-identify by comparing the anonymized data with other data sources. We anonymize the data using k-anonymity [6] as the data utility is essential. k-anonymity is a technique developed to mitigate the possibility of anonymized data being re-identified when connecting to other datasets. k-anonymity helps to balance data utility and privacy. It is achieved by generalization and suppression.

The metadata in the module builds a layer of integration called the metadata layer above the local databases. The most common integration system uses a mediator and wrapper for integration. Mediator-Wrapper has several issues like Scalability, Flexibility, and Adaptability [56]. However, we build metadata or a semantic layer.

The metadata layer helps users independently view data using common words. It maps nuanced data into common words such as patient, age, or city to provide an organization-wide coherent, condensed view of the data. It is not a specific abstraction. It is a collection of abstractions that are used to resolve multiple issues.

We consider three databases for the experimental scenario: local DB1, local DB2, and local DB3 from data source1, data source2, and data source3. The local DBs are a collection of tables and records that we need to bring together in one database, the integrated database. The integrated database contains all the tables that make up each local DBs. The integration layer is the layer of metadata that includes information that defines the three databases. The information stored in the metadata layer can be listed as follows.

- Canonical representation: It corresponds to an object's name at the metadata layer. It is the global name of the object.
- Data type and semantic difference: It is a definition of the local data types in each database used for each column name.
- Supplementary Fields: It stores additional mapping or conversion information between canonical to local DB names.

3.1.3 Data Aggregation Module

The main task of the module involves integrating data from multiple databases into a coherent, cohesive view. In a standard data integration method, the user sends a query to the central server. The central server then collects the data needed from diverse sources. The data is derived from the origins and compiled into a unified, coherent set of data. Finally, it is fetched to the requested user.

3.2 Framework Flow

In our application scenario, we started with the user requesting access to the system. The user access request goes through a comprehensive check based on the user attribute check with the knowledge graph. Then the data shared by the sources were prepared by removing sensitive attributes and anonymizing using k-anonymity. After that, we built a metadata layer based on the shared databases. We now have the integration layer on top of the participating databases from the previous stage. There are two layers, the Global layer and the local layer. The global layer has metadata information and canonical representation. The local layer holds the local databases.

The user sends a query using the canonical names of fields in the metadata layer. The global query is broken down into subqueries, one for each participating local database. The subqueries are sent and executed by the applicable local databases. The outcomes are put together using a UNION process.

3.3 Use Cases

Data sharing and integration are widespread every day throughout organizations. Many organizations or tasks benefit from getting data integrated from different sources. We describe here two use cases that can benefit from our system.

1. Disease Outbreak: Early identification of infectious diseases is critical in avoiding life-threatening contagious diseases. Virulent disease attacks such as SARS and bird flu have rendered disease monitoring a significant issue. Outbreak identification performs well when integrating and analyzing several data points in real time. Safeguarding identity disclosure by appropriate privacy-preserving data aggregation and sharing strategies would enable healthcare advancements.
2. Healthcare Research: Analysis of disease occurrence, frequency, and clinical signs is essential to identifying and managing them. These analyzes have an enormous effect on government policies. An apparent precondition for this research is the availability of the necessary data. The data then must be gathered and integrated by various healthcare facilities while sanitizing information that is vulnerable to privacy. Concerns over privacy are a significant barrier to simplifying such activities. A lack of privacy can cause serious harm to participants. Another issue with the violation of privacy is the chances of bias from apparently definitive statistical findings against different sub-groups. Often, privacy is handled by avoiding sharing rather than incorporating data processing restrictions into the mechanism. The privacy-preserving integration and sharing of research data in healthcare have become critical to encouraging scientific exploration.

4 Dataset Description

We used the MIMIC-III dataset [30] to produce a synthetic data set for three independent sources. We assume each source represents a hospital. Each data source has more than one thousand patient records. The number of attributes in each source is variable because some organizations may want to store more features to know their patient better. In contrast, others might prefer to store minimal information. We have eighteen attributes, like name, age, gender, Social Security Number (SSN), city, religion, insurance, diagnosis, drug, medical conditions, admission type, etc., from source one. Likewise, we have ten attributes from source two and twelve from source three. The data that are available from the sources are confidential and sensitive. If either of the entities shares the way-it-is info, this may lead to a breach of privacy.

5 Evaluation

We developed a proof of concept to evaluate our framework. Let's say a team of researchers at the National Insitute of Health (NIH) would like to know about a few statistics related to COVID-19 cases. The team comprises several individuals. There is a senior research scientist, a few junior scientists, and many interns. The team would like to know the number of positive covid cases in each county in California. The team wants to see the patient's age groups, medical conditions, etc. To achieve this, the team asks for data from hospitals. The hospitals then agree to share the data by removing the sensitive or identifier attributes and anonymizing the data using k-anonymity. The team creates a knowledge graph and defines the user attribute in the knowledge graph along with their access pattern. The team also creates a metadata layer on top of the data shared by the hospitals. An intern has been asked to report California's average cases per county. Once an intern makes a request into the framework, the attributes get checked with the knowledge graph. After that, the intern puts a query that gets written into subqueries and runs into data sources. The query result is combined from all the sources and displayed. If the senior researcher in the team performs the same operation, the final results contain more rows and columns as the attributes allow to get access to complete data. Likewise, a junior researcher gets more rows and columns than the intern.

6 Conclusion

This paper proposed a framework for privacy-preserving data sharing and integration. The framework used a knowledge graph for providing attribute-based access control. The access decisions were evaluated comprehensively based on the rules defined in the knowledge graph. The data shared was anonymized using k-anonymity. A semantic layer was created to represent the schema mappings and then populated with the metadata records. Later, data was integrated from different Local DBs based on the user query. The framework helps to share data and integrate securely, protecting individual privacy. The framework can be used for several use cases like to study a disease outbreak.

There can be several expansions for this work. One of the challenging aspects would be to include many data sources and then check the system's performance. Obviously, with the addition of more sources, the complexities would add up in the metadata layer. Exploring and simplifying the challenges in building the metadata layer would be interesting. Often, machine learning is used these days for schema mapping. It would be a great area to explore. Another promising research direction would be incorporating access control into the integrated document's fields. Often, EHRs have this field-level access control. Other data anonymizing techniques could be explored as the data is very confidential, and at the same time, the utility is essential.

Acknowledgements This work has been supported by the Center for Accelerated Real Time Analytics (CARTA), Office of Naval Research (ONR) under grants N00014-18-1-2453, N00014-19-WX-00568, and N00014-20-WX01704 and National Science Foundation (NSF) under grant 1955319.

References

1. Halevy A, Rajaraman A, Ordille J (2006) Data integration: the teenage years. In: Proceedings of the 32nd international conference on Very large data bases, pp 9–16
2. English A, Ford CA (2004) The hipaa privacy rule and adolescents: legal questions and clinical challenges. Perspect Sexual Reproduct Health 36(2):80–86
3. Sharma S (2019) Data privacy and GDPR handbook. Wiley
4. McGuinness DL, Van Harmelen F et al. (2004) Owl web ontology language overview. W3C Recommend 10(10):2004
5. Hu VC, Ferraiolo D, Kuhn R, Friedman AR, Lang AJ, Cogdell MM, Schnitzer A, Sandlin K, Miller R, Scarfone K et al (2013) Guide to attribute based access control (abac) definition and considerations (draft). NIST Spec Publicat 800(162):1–54
6. Latanya S (2002) k-anonymity: a model for protecting privacy. Int J Uncertain, Fuzziness Knowl-Based Syst 10(05):557–570
7. Bahga A, Madisetti VK (2015) Healthcare data integration and informatics in the cloud. Computer 48(2):50–57
8. Clifton C, Kantarcioğlu M, Doan A, Schadow G, Vaidya J, Elmagarmid A, Suciu D (2004) Privacy-preserving data integration and sharing. In: Proceedings of the 9th ACM SIGMOD workshop on research issues in data mining and knowledge discovery, pp 19–26
9. Bhowmick SS, Gruenwald L, Iwaihara M, Chatvichienchai S (2006) Private-iye: a framework for privacy preserving data integration. In: 22nd international conference on data engineering workshops (ICDEW'06). IEEE, p 91
10. Lu Y, Sinnott RO (2020) Security and privacy solutions for smart healthcare systems. In: Innovation in health informatics. Elsevier, pp 189–216
11. Yau SS, Yin Y (2008) A privacy preserving repository for data integration across data sharing services. IEEE Trans Servic Comput 1(3):130–140
12. Agrawal R, Kiernan J, Srikant R, Xu Y (2002) Hippocratic databases. In: Proceedings of the 28th international conference very large databases (VLDB)
13. Cranor L, Langheinrich M, Marchiori M, Presler-Marshall M, Reagle J (2002) The platform for privacy preferences 1.0 (p3p1. 0) specification. W3C Recommend 16
14. van den Braak SW, Choenni S, Meijer R, Zuiderwijk A (2012) Trusted third parties for secure and privacy-preserving data integration and sharing in the public sector. In: Proceedings of the 13th annual international conference on digital government research, pp 135–144
15. Lindell Y, Pinkas B (2000) Privacy preserving data mining. In: Advances in cryptology-CRYPTO 2000: 20th annual international cryptology conference Santa Barbara, California, USA, August 20–24, 2000 Proceedings. Springer, pp 36–54
16. Du W, Atallah MJ (2001) Secure multi-party computation problems and their applications: a review and open problems. In: Proceedings of the 2001 workshop on New security paradigms, pp 13–22
17. Murat K, Chris C (2004) Privacy-preserving distributed mining of association rules on horizontally partitioned data. IEEE Trans Knowl Data Engin 16(9):1026–1037
18. Agrawal R, Srikant R (2000) Privacy-preserving data mining. In: Proceedings of the 2000 ACM SIGMOD international conference on management of data, pp 439–450
19. Kargupta H, Datta S, Wang Q, Sivakumar K (2003) On the privacy preserving properties of random data perturbation techniques. In: Third IEEE international conference on data mining. IEEE, pp 99–106

20. Searls DB (2005) Data integration: challenges for drug discovery. Nat Rev Drug Discov 4(1):45–58
21. David G-C, Imad A, Dieter M, Andrew T, Matthias M, Andreas G, Esteban B, Erik B-R, Ana C, Jesper T (2014) Data integration in the era of omics: current and future challenges. BMC Syst Biol 8(2):1–10
22. Tatbul N (2010) Streaming data integration: challenges and opportunities. In: 2010 IEEE 26th international conference on data engineering workshops (ICDEW 2010). IEEE, pp 155–158
23. Vladimir G, Nataša P (2015) Methods for biological data integration: perspectives and challenges. J R Soc Interface 12(112):20150571
24. Sarmad A, Farouk M, Tillal E (2003) Healthcare information management: the integration of patients' data. Logist Inf Manag 16(3/4):286–295
25. David B, Michael R, Pearl B, Mark T (2007) A data integration broker for healthcare systems. Computer 40(4):34–41
26. Zheleva E, Getoor L (2011) Privacy in social networks: a survey. Soc Netw Data Analyt 277–306
27. Dwork C (2008) Differential privacy: a survey of results. In: Theory and applications of models of computation: 5th international conference, TAMC 2008, Xi'an, China, April 25–29, 2008. Proceedings, vol 5. Springer, pp 1–19
28. Yuan E, Tong J (2005) Attributed based access control (abac) for web services. In: IEEE international conference on web services (ICWS'05). IEEE
29. Joshi KP, Yesha Y, Finin T et al. (2016) An ontology for a hipaa compliant cloud service. In: 4th international ibm cloud academy conference ICACON 2016
30. Johnson AEW, Pollard TJ, Shen L, Lehman L-WH, Feng M, Ghassemi M, Moody B, Szolovits P, Celi LA, Mark RG (2016) Mimic-iii, a freely accessible critical care database. Scientific Data 3(1):1–9

Equivariant LS-Category and Topological Complexity of Product of Several Manifolds

R. Karthika◉ and V. Renukadevi◉

Abstract The LS-category and the topological complexity are some homotopy invariants of a topological space, and the topological complexity is a close relative of the LS-category. In this paper, we discuss some $\mathbb{Z}_2$-spaces, and we find the equivariant version of LS-category and topological complexity of these spaces.

Keywords Equivariant LS-category · Equivariant topological complexity · $\mathbb{Z}_2$-space · Generalized Dold manifolds

1 Introduction

The concept of Lusternik-Schnirelmann category (LS-category) was introduced by Lusternik and Schnirelmann [12] which is the smallest integer k, such that X may be covered by k open subsets $V_1, V_2, \ldots, V_k$ of X with each inclusion $V_i \hookrightarrow X$ being null-homotopic and denoted by $cat(X)$. If no such k exists, we will set $cat(X) = \infty$. And it is bounded below by $cl_R(X) + 1$ [3, Proposition 1.5]. The concept of topological complexity was introduced by Farber [8] to study motion planning in mechanical systems. Topological complexity of a path connected space X is a homotopy invariant which is defined by the minimal integer k, such that the Cartesian product $X \times X$ may be covered by k open subsets $U_1, U_2, \ldots, U_k$ such that for any $i = 1, 2, \ldots, k$, there exists a continuous motion planning $s_i : U_i \to PX$, $\pi \circ s_i = id$ over U_i. If no such k exists, then $TC(X) = \infty$. And it is bounded below by $zcl_R(X) + 1$ [8, Theorem 7]. One can estimate the bound for topological complexity by using LS-category.

The authors thank the UGC-CSIR(File No.:09/1327(13699)/2022-EMR-I), New Delhi for providing the financial support.

R. Karthika (✉) · V. Renukadevi
Department of Mathematics, School of Mathematics and Computer Sciences, Central University of Tamil Nadu, Thiruvarur 610 005, India
e-mail: karthikar5699@gmail.com

In this paper, we discuss the equivariant versions of these homotopy invariants. Marzantowicz [13] introduced the equivariant LS-category, denoted by $cat_G(X)$ for a G-space X, where G is a compact Hausdorff space. Colman and Grant [2] introduced the concept of equivariant topological complexity. Suppose the group action has nonempty fixed point set, then we can estimate the bound for equivariant topological complexity by using equivariant LS-category.

In Sect. 2, we give examples of $\mathbb{Z}_2$-spaces. In Sects. 3 and 4, we discuss the properties of equivariant LS-category and topological complexity, after that we calculate the equivariant LS-category and topological complexity of $\mathbb{Z}_2$-spaces which are given in Sect. 2.

2 G-Space

Definition 1 Let G be a compact Hausdorff topological group and X be a Hausdorff space. Suppose G is acting continuously on X, then X is called G-space. The set $Gx = \{g.x | g \in G\}$ is called orbit of x and it is denoted by $O(x)$.

Definition 2 If H is a closed subgroup of G, then $X^H = \{x \in X | hx = x$ for all $h \in H\}$ is called the H-fixed point set of X.

Definition 3 Let X and Y be G-spaces and $\phi, \psi : X \to Y$ be two G-maps. Suppose there is a G-map $F : X \times I \to Y$ with $F(x, 0) = \phi(x)$ and $F(x, 1) = \psi(x)$, where G acts trivially on I and diagonally on $X \times I$. Then ϕ and ψ are said to be G-homotopic.

Fix $G = \mathbb{Z}_2$ and consider the following.

Example 1 Consider an m-dimensional manifold M and an n-dimensional manifold N with an involution $\tau : M \to M$ and $\sigma : N \to N$ such that σ has a non-empty fixed point set. Define $\psi : M \times N \to M \times N$ by $\psi(x, y) = (\tau(x), \sigma(y))$. Now denote the orbit space
$$X(M, N) = \frac{M \times N}{(x, y) \sim (\tau(x), \sigma(y))}.$$
Then $X(M, N)$ is the generalized Dold space with dimension $m + n$ [17]. And also note that Dold manifolds [6], projective product space [5], and the generalized Dold manifolds [16] are all examples of this class of manifolds. Suppose τ is a fixed point-free involution, then the involution ψ freely acts on $M \times N$. This gives a free $\mathbb{Z}_2$-action on $M \times N$.

Example 2 Let $1 \le i \le r$, n_i be positive integers, define
$$P(n_1, n_2, \ldots, n_r) = S^{n_1} \times S^{n_2} \times \cdots \times S^{n_r} / ((x_1, x_2, \ldots, x_r) \sim (-x_1, -x_2, \ldots, -x_r)),$$
where $x_i = S^{n_i}$. This is a manifold of dimension $n_1 + n_2 + \cdots + n_r$, which we call a projective product space [5]. This gives the fixed point free $\mathbb{Z}_2$-action on $S^{n_1} \times S^{n_2} \times \cdots \times S^{n_r}$.

Example 3 Consider the smooth manifold M and the space $M \times \mathbb{C}P^n$. If $\tau : M \to M$ is an involution and $\sigma : \mathbb{C}P^n \to \mathbb{C}P^n$ is a complex conjugation involution, then the involution $\psi : M \times \mathbb{C}P^n \to M \times \mathbb{C}P^n$ defined by $\psi(x, y) = (\tau(x), \sigma(y))$ gives the $\mathbb{Z}_2$-action on $M \times \mathbb{C}P^n$. We denote the orbit space $X(M, \mathbb{C}P^n) = \frac{M \times \mathbb{C}P^n}{(x,y) \sim (\tau(x), \sigma(y))}$. This induces a fiber bundle $\mathbb{C}P^n \hookrightarrow X(M, \mathbb{C}P^n) \xrightarrow{p} \frac{M}{\tau}$. Suppose τ is a fixed point free involution, then the involution ψ freely acts on $M \times \mathbb{C}P^n$. This gives a free $\mathbb{Z}_2$-action on $M \times \mathbb{C}P^n$.

Consider $1 \leq m \leq n - 1$ and $Gr_m(\mathbb{C}^n)$ denotes the set of all m-dimensional subspace in $\mathbb{C}^n$. The space $Gr_m(\mathbb{C}^n)$ is known as *complex Grassmann manifold* [14] and it has complex dimension $m(n - m)$ (real dimension of $Gr_m(\mathbb{C}^n)$ is $2m(n - m)$). In particular, $Gr_1(\mathbb{C}^{n+1}) = \mathbb{C}P^n$ is the complex projective plane. A m-tuple $\lambda = (\lambda_1, \ldots, \lambda_m)$ is called a *Schubert symbol* if $1 \leq \lambda_1 < \cdots < \lambda_m \leq n$. Consider $\mathbb{C}^l := \{(z_1, \ldots, z_l, 0, \ldots, 0) \in \mathbb{C}^n\}$. Now define the a *Schubert cell* $E(\lambda)$ for the Schubert symbol λ as follows:
$E(\lambda) = \{V \in Gr_m(\mathbb{C}^n) | \dim(V \cap Gr_m(\mathbb{C}^{\lambda_j})) = j, \dim(V \cap Gr_m(\mathbb{C}^{\lambda_j - 1})) = j - 1$ for $j = 1, 2, \ldots, m\}$. It is clear that $E(\lambda)$ is of even dimension and it gives the cell structure on $Gr_m(\mathbb{C}^n)$ [14]. It is well known that the cup length of $Gr_m(\mathbb{C}^n)$ is $m(n - m)$.

Example 4 Consider the smooth manifold M and the space $M \times Gr_m(\mathbb{C}^n)$. If $\tau : M \to M$ is an involution and $\sigma : Gr_m(\mathbb{C}^n) \to Gr_m(\mathbb{C}^n)$ is a complex conjugation involution, then the involution $\psi : M \times Gr_m(\mathbb{C}^n) \to M \times Gr_m(\mathbb{C}^n)$ defined by $\psi(x, y) = (\tau(x), \sigma(y))$ gives the $\mathbb{Z}_2$-action on $M \times Gr_m(\mathbb{C}^n)$. We denote the orbit space $X(M, Gr_m(\mathbb{C}^n)) = \frac{M \times Gr_m(\mathbb{C}^n)}{(x,y) \sim (\tau(x), \sigma(y))}$. This induces a fiber bundle $Gr_m(\mathbb{C}^n) \hookrightarrow X(M, Gr_m(\mathbb{C}^n)) \xrightarrow{p} \frac{M}{\tau}$. Suppose τ is a fixed point free involution, then the involution ψ freely acts on $M \times Gr_m(\mathbb{C}^n)$. This induces a free $\mathbb{Z}_2$-action on $M \times Gr_m(\mathbb{C}^n)$.

Consider the increasing sequence of subspaces of $\mathbb{C}^n$ as $\{0\} = V_0 \subset V_1 \subset V_2 \subset \cdots \subset V_n = \mathbb{C}^n$, where $\dim(V_i) = i$. This is called the *complete flag* on $\mathbb{C}^n$. Denote $Fl(n) = \{V_\bullet = (\{0\} = V_0 \subset V_1 \subset V_2 \subset \cdots \subset V_n = \mathbb{C}^n)\}$ as the set of all complete flags on $\mathbb{C}^n$. Here $Fl(n)$ is known as *complete flag manifold* with complex dimension $\frac{n(n-1)}{2}$ (real dimension of $Fl(n)$ is $n(n - 1)$). Let $\{e_1, \ldots, e_n\}$ be the standard basis of $\mathbb{C}^n$. Let $F_\bullet = (\{0\} = F_0 \subset F_1 \subset F_2 \subset \cdots \subset F_n = \mathbb{C}^n)$, where $F_i = \{e_1, \ldots e_i\}$ is the standard complete flag of $\mathbb{C}^n$. Consider the symmetric group S_n. For each $\omega \in S_n$, define $E(\omega) = \{V_\bullet \in Fl(n) | \dim(V_p \cap F_q) = \#\{i \leq p | \omega(i) \leq q\}$, for every $1 \leq p, q \leq n\}$. Then $E(\omega)$ is an open cell of real dimension $2l(\omega)$, where $l(\omega) = \#\{i < j | \omega(i) < \omega(j)\}$. For each $\omega \in S_n$, we have $0 \leq l(\omega) \leq \frac{n(n-1)}{2}$. This gives the cell structure on $Fl(n)$ [10]. It is well known that the cup length of $Fl(n)$ is $\frac{n(n-1)}{2}$.

Example 5 Consider the smooth manifold M and the space $M \times Fl(n)$. If $\tau : M \to M$ is an involution and $\sigma : Fl(n) \to Fl(n)$ is a complex conjugation involution, then the involution $\psi : M \times Fl(n) \to M \times Fl(n)$ defined by $\psi(x, y) = (\tau(x), \sigma(y))$ gives the $\mathbb{Z}_2$-action on $M \times Fl(n)$. We denote the orbit space $X(M, Fl(n)) = \frac{M \times Fl(n)}{(x,y) \sim (\tau(x), \sigma(y))}$. This induces a fiber bundle $Fl(n) \hookrightarrow X(M, Fl(n)) \xrightarrow{p} \frac{M}{\tau}$. Suppose

τ is a fixed point free involution, then the involution ψ freely acts on $M \times Fl(n)$. This gives a free $\mathbb{Z}_2$-action on $M \times Fl(n)$.

3 Properties of Equivariant LS-Category

In this section, we recall the definition and properties of the equivariant LS-category of a G-space. After that we calculate the equivariant LS-category of $\mathbb{Z}_2$-spaces given in the previous section.

Definition 4 Let X be a G-space and U be an invariant subset of X. Suppose the inclusion $i_U : U \to X$ is G-homotopic to a map with values in a single orbit, then U is called G-categorical.

Definition 5 [13] The equivariant category of a G-space X, denoted by $cat_G(X)$, is the least integer k such that X may be covered by k open sets $U_1, \ldots, U_k$, each of U_i being G-categorical.

Theorem 1 [7, 13] *When X is a free metrizable G-space, we have $cat_G(X) = cat(X/G)$, where X/G is the orbit space of the corresponding group action.*

Lemma 1 [2, Theorem 3.15] *Let X and Y be G-connected G-spaces such that $X \times Y$ are completely normal. If $X^G \neq \emptyset$ or $Y^G \neq \emptyset$, then $cat_G(X \times Y) \leq cat_G(X) + cat_G(Y) - 1$, where $X \times Y$ is given the diagonal G-action.*

Theorem 2 *Consider $\mathbb{C}P^r$ and $Gr_m(\mathbb{C}^n)$ with the $\mathbb{Z}_2$-action, and $\tau : \mathbb{C}P^r \to \mathbb{C}P^r$ and $\sigma : Gr_m(\mathbb{C}^n) \to Gr_m(\mathbb{C}^n)$ be a complex conjugation involution. Then $cat_{\mathbb{Z}_2}(\mathbb{C}P^r \times Gr_m(\mathbb{C}^n)) = r + m(n - m) + 1$, where $\mathbb{C}P^r \times Gr_m(\mathbb{C}^n)$ is given the diagonal $\mathbb{Z}_2$-action.*

Proof The $\mathbb{Z}_2$-action on $\mathbb{C}P^r$ and $Gr_m(\mathbb{C}^n)$ implies that $(\mathbb{C}P^r)^{\mathbb{Z}_2} = \mathbb{R}P^r$, $(Gr_m(\mathbb{C}^n))^{\mathbb{Z}_2} = Gr_m(\mathbb{R}^n)$, and $(\mathbb{C}P^r \times Gr_m(\mathbb{C}^n))^{\mathbb{Z}_2} = \mathbb{R}P^r \times Gr_m(\mathbb{R}^n)$. Note that $cat_{\mathbb{Z}_2}(\mathbb{C}P^r) = r + 1$ [15, Proposition 3.10] and $cat_{\mathbb{Z}_2}(Gr_m(\mathbb{C}^n)) = m(n - m) + 1$ [4, Theorem 5.1]. By Lemma 2.20 in [1] and Lemma 1, we have $r + m(n - m) + 1 \leq cat_{\mathbb{Z}_2}(\mathbb{C}P^r \times Gr_m(\mathbb{C}^n)) \leq r + m(n - m) + 1$. Similarly, we can calculate the following:
$cat_{\mathbb{Z}_2}(Gr_m(\mathbb{C}^n) \times Fl(k)) = m(n - m) + \frac{k(k-1)}{2} + 1$ and
$cat_{\mathbb{Z}_2}(\mathbb{C}P^r \times Fl(k)) = r + \frac{k(k-1)}{2} + 1$, where $\mathbb{Z}_2$-action on $Gr_m(\mathbb{C}^n)$, $\mathbb{C}P^r$, and $Fl(k)$ is a complex conjugation action.

Theorem 3 *Let M be a simply connected metrizable space with a free involution τ and $\mathbb{C}P^n$ with conjugation involution σ and the $\mathbb{Z}_2$-action ψ on $M \times \mathbb{C}P^n$ defined as in Example 3. Then $n + cl_{\mathbb{Z}_2}(M/\tau) + 1 \leq cat_{\mathbb{Z}_2}(M \times \mathbb{C}P^n) \leq n + cat(M/\tau)$.*

Proof Since M is a metrizable space, $M \times \mathbb{C}P^n$ is also metrizable and ψ is a free $\mathbb{Z}_2$-action, so we have $cat_{\mathbb{Z}_2}(M \times \mathbb{C}P^n) = cat(X(M, \mathbb{C}P^n))$. Therefore, the result follows from Theorem 5.3 in [4].

Theorem 4 *Let M be a simply connected metrizable space with a free invo-lution τ and $Gr_m(\mathbb{C}^n)$ with conjugation involution σ and the $\mathbb{Z}_2$-action ψ on $M \times Gr_m(\mathbb{C}^n)$ defined as in Example 4. Then $n + cl_{\mathbb{Z}_2}(M/\tau) + 1 \leq cat_{\mathbb{Z}_2}(M \times Gr_m(\mathbb{C}^n)) \leq m(n - m) + cat(M/\tau)$.*

Proof Since M is a metrizable space, $M \times \mathbb{C}P^n$ is also metrizable and ψ is a free $\mathbb{Z}_2$-action, so we have $cat_{\mathbb{Z}_2}(M \times Gr_m(\mathbb{C}^n)) = cat(X(M, Gr_m(\mathbb{C}^n)))$. By Theorem 4.11 in [4], we get that the cup length of $X(M, Gr_m(\mathbb{C}^n)) \geq m(n - m) + cl_{\mathbb{Z}_2}(M/\tau)$. From Theorem 5.1 of [4], we have $cat_{\mathbb{Z}_2}(Gr_m(\mathbb{C}^n)) = m(n - m) + 1$. Therefore, the right side inequality follows from Theorem 2.6 in [15].

Theorem 5 *Let M be a simply connected metrizable space with a free involution τ and $Fl(n)$ with conjugation involution σ and the $\mathbb{Z}_2$-action ψ on $M \times Fl(n)$ defined as in Example 5. Then $\frac{n(n-1)}{2} + cl_{\mathbb{Z}_2}(M/\tau) + 1 \leq cat_{\mathbb{Z}_2}(M \times Fl(n)) \leq \frac{n(n-1)}{2} + cat(M/\tau)$.*

Proof Since M is a metrizable space, $M \times Fl(n)$ also metrizable and ψ is a free $\mathbb{Z}_2$-action, so we have $cat_{\mathbb{Z}_2}(M \times Fl(n)) = cat(X(M, Fl(n)))$. By Theorem 4.11 in [4], we get that the cup length of $X(M, Fl(n))) \geq \frac{n(n-1)}{2} + cl_{\mathbb{Z}_2}(M/\tau)$. Let $0 \leq i \leq \frac{n(n-1)}{2}$, and consider $U_i = \bigcup_{l(\omega) \leq i} E(\omega)$. Then for each i, U_i is a subcomplex of $Fl(n)$. Therefore, for each i, there exists a conjugation invariant open neighborhood V_i of U_i such that V_i retracts on U_i. Let $V_{-1} = \emptyset$, then $\{V_i - V_{i-1}\}_{i=0}^{\frac{n(n-1)}{2}}$ is a conjugation invariant categorical cover of $Fl(n)$. This implies that $cat_{\mathbb{Z}_2}(Fl(n)) = \frac{n(n-1)}{2} + 1$. Therefore, $cat_{\mathbb{Z}_2}(M \times Fl(n)) \leq \frac{n(n-1)}{2} + cat(M/\tau)$, by Theorem 2.6 in [15].

Corollary 1 *Let $M = S^{n_1} \times S^{n_2} \times \cdots \times S^{n_r}$ with antipodal action τ, and $Fl(n)$ with conjugation involution. Then $cat_{\mathbb{Z}_2}(S^{n_1} \times S^{n_2} \times \cdots \times S^{n_r} \times Fl(n)) = \frac{n(n-1)}{2} + n_1 + r$.*

Proof The cohomology of the projective product space [5] yields the cup length of $S^{n_1} \times S^{n_2} \times \cdots \times S^{n_r}/\tau$ is $n_1 + r - 1$ and $cat(S^{n_1} \times S^{n_2} \times \cdots \times S^{n_r}/\tau) = n_1 + r$ [9, Theorem 1.2]. , by Theorem 5, we get that $cat_{\mathbb{Z}_2}(S^{n_1} \times S^{n_2} \times \cdots \times S^{n_r} \times Fl(n)) = \frac{n(n-1)}{2} + n_1 + r$.

4 Properties of Equivariant Topological Complexity

In this section, we recall the definition and properties of the equivariant topological complexity of a G-space. And further we calculate the equivariant topological complexity of a $\mathbb{Z}_2$-space given in Sect. 2.

Definition 6 The equivariant topological complexity of a G-space X is the least integer k such that $X \times X$ is covered by k-invariant open sets $U_1, \ldots, U_k$, on each of which there is a G-equivariant map $s_i : U_i \to X^I$ such that $\phi \circ s_i \simeq_G iU_i$, and it is denoted by $TC_G(X)$. If no such integer exists, then we set $TC_G(X) = \infty$. Observe that $TC(X) \leq TC_G(X)$.

Theorem 6 [11, Theorem 4.2] *Let G be a compact Lie group, and let X, and Y be smooth G-manifolds. Then $TC_G(X \times Y) \leq TC_G(X) + TC_G(Y) - 1$, where $X \times Y$ is given the diagonal G-action.*

Theorem 7 *Let G be a compact Lie group, $\mathbb{K}$ be an infinite field and X_i be smooth manifold with $TC_G(X_i) = 2n_i + 1$, for all $1 \leq i \leq n$. Suppose for each i, there exists $[w_i] \in H^*(X_i, \mathbb{K})$ such that $([w_i] \otimes 1 - 1 \otimes [w_i])^{2n_i} \neq 0$, then $TC_G(X_1 \times X_2 \times \cdots \times X_k) = \sum_{i=1}^{k} 2n_i + 1$, where $(X_1 \times X_2 \times \cdots \times X_k)$ is given the diagonal G-action.*

Proof For each i, $([w_i] \otimes 1 - 1 \otimes [w_i])^{2n_i} = (-1)^n \binom{2n_i}{n_i} \cdot [w]^{n_i} \otimes [w]^{n_i} \neq 0$. By Künneth formula, we have
$H^*(X_1 \times X_2 \times \cdots \times X_k, \mathbb{K}) = H^*(X_1, \mathbb{K}) \otimes \cdots \otimes H^*(X_k, \mathbb{K})$. Set

$$a_1 = ([w_1] \otimes 1 \cdots \otimes 1) \otimes (1 \otimes \cdots \otimes 1) - (1 \otimes \cdots \otimes 1) \otimes ([w_1] \otimes 1 \cdots \otimes 1)$$
$$a_2 = (1 \otimes [w_2] \cdots \otimes 1) \otimes (1 \otimes \cdots \otimes 1) - (1 \otimes \cdots \otimes 1) \otimes (1 \otimes [w_2] \cdots \otimes 1)$$
$$\vdots$$
$$a_k = (1 \otimes 1 \cdots \otimes [w_k]) \otimes (1 \otimes \cdots \otimes 1) - (1 \otimes \cdots \otimes 1) \otimes (1 \otimes 1 \cdots \otimes [w_k])$$

Note that for each $i \in \{1, 2, \ldots k\}$, we have
$a_i \in (H^*(X_1, \mathbb{K}) \otimes \cdots \otimes H^*(X_k, \mathbb{K})) \otimes (H^*(X_1, \mathbb{K}) \otimes \cdots \otimes H^*(X_k, \mathbb{K}))$ such
that $a_i^{2n_i} \neq 0$. From this, we have $zcl_{\mathbb{K}}(X_1 \times X_2 \times \cdots \times X_k) \geq \sum_{i=1}^{k} 2n_i$. This implies
that $TC(X_1 \times X_2 \times \cdots \times X_k) \geq \sum_{i=1}^{k} 2n_i + 1$ [8, Theorem 7]. Since for any topological space X, $TC(X) \leq TC_{\mathbb{Z}_2}(X)$, we have $TC_{\mathbb{Z}_2}(X_1 \times X_2 \times \cdots \times X_k) = \sum_{i=1}^{k} 2n_i + 1$.

Corollary 2 *Consider the smooth manifold X with $TC_{\mathbb{Z}_2}(X) = 2n + 1$. Suppose there exists $[w] \in H^*(X, \mathbb{K})$ such that $([w] \otimes 1 - 1 \otimes [w])^{2n} \neq 0$, then $TC_{\mathbb{Z}_2}(X^n) = 2kn + 1$, where $X^n = \underbrace{X \times X \times \cdots \times X}_{n\text{-times}}$ is given the diagonal G-action.*

Proof The proof follows from Theorem 7.

Theorem 8 *Let M be a smooth $\mathbb{Z}_2$-manifold with an involution τ and $\mathbb{C}P^n$ with complex conjugation action. Then from the $\mathbb{Z}_2$-action determined by Example 3, we have $2n + zcl_{\mathbb{Q}}(M) + 1 \leq TC_{\mathbb{Z}_2}(M \times \mathbb{C}P^n) \leq 2n + TC_{\mathbb{Z}_2}(M)$.*

Proof By Künneth formula, we have
$H^*(M \times \mathbb{C}P^n, \mathbb{Q}) = H^*(M, \mathbb{Q}) \times H^*(\mathbb{C}P^n, \mathbb{Q})$.
Therefore, $zcl_{\mathbb{Q}}(M \times \mathbb{C}P^n) = zcl_{\mathbb{Q}}(M) + zcl_{\mathbb{Q}}(\mathbb{C}P^n) = 2n + 1 + zcl_{\mathbb{Q}}(M)$. By Theorem 7 in [8], we have $2n + 1 + zcl_{\mathbb{Q}}(M) \leq TC(M \times \mathbb{C}P^n) \leq TC_{\mathbb{Z}_2}(M \times \mathbb{C}P^n)$. The right side inequality follows from Theorem 6 and Example 6.6 in [4].

Theorem 9 *Let M be a smooth $\mathbb{Z}_2$-manifold with an involution τ and $Gr_m(\mathbb{C}^n)$ with complex conjugation action. Then from the $\mathbb{Z}_2$-action determined by Example 4, we have $2m(n-m) + zcl_{\mathbb{Q}}(M) + 1 \leq TC_{\mathbb{Z}_2}(M \times Gr_m(\mathbb{C}^n)) \leq 2n + TC_{\mathbb{Z}_2}(M)$.*

Proof By Künneth formula, we have
$H^*(M \times Gr_m(\mathbb{C}^n), \mathbb{Q}) = H^*(M, \mathbb{Q}) \times H^*(Gr_m(\mathbb{C}^n), \mathbb{Q})$.
Therefore, $zcl_{\mathbb{Q}}(M \times Gr_m(\mathbb{C}^n)) = zcl_{\mathbb{Q}}(M) + zcl_{\mathbb{Q}}(Gr_m(\mathbb{C}^n)) = 2m(n-m) + 1 + zcl_{\mathbb{Q}}(M)$. By Theorem 7 in [8], we have $2n + 1 + zcl_{\mathbb{Q}}(M) \leq TC(M \times Gr_m(\mathbb{C}^n)) \leq TC_{\mathbb{Z}_2}(M \times Gr_m(\mathbb{C}^n))$. Therefore, $TC_{\mathbb{Z}_2}(M \times Gr_m(\mathbb{C}^n)) \leq 2n + TC_{\mathbb{Z}_2}(M)$, by Theorem 6 and Example 6.6 in [4].

Theorem 10 *Let M be a smooth $\mathbb{Z}_2$-manifold with an involution τ and $Fl(n)$ with complex conjugation action. Then from the $\mathbb{Z}_2$-action determined by Example 5, we have $n(n-1) + zcl_{\mathbb{Q}}(M) + 1 \leq TC_{\mathbb{Z}_2}(M \times Fl(n)) \leq n(n-1) + TC_{\mathbb{Z}_2}(M)$.*

Proof By Künneth formula, we have
$H^*(M \times Fl(n), \mathbb{Q}) = H^*(M, \mathbb{Q}) \times H^*(Fl(n), \mathbb{Q})$.
Therefore, $zcl_{\mathbb{Q}}(M \times \mathbb{C}P^n) = zcl_{\mathbb{Q}}(M) + zcl_{\mathbb{Q}}(Fl(n)) = n(n-1) + 1 + zcl_{\mathbb{Q}}(M)$. By Theorem 7 in [8], we have $n(n-1) + 1 + zcl_{\mathbb{Q}}(M) \leq TC(M \times \mathbb{C}P^n) \leq TC_{\mathbb{Z}_2}(M \times Fl(n))$. From Theorem 5, we have $cat_{\mathbb{Z}_2}(Fl(n)) = \frac{n(n-1)}{2} + 1$. Therefore $TC_{\mathbb{Z}_2}(Fl(n)) \leq n(n-1) + 1$ [2, Corollary 5.8]. Since $zcl_{\mathbb{Q}}(Fl(n)) = n(n-1)$, we have $TC_{\mathbb{Z}_2}(Fl(n)) = n(n-1) + 1$. Therefore, the right side inequality follows from Theorem 6.

Corollary 3 *Let $M = S^{n_1} \times S^{n_2} \times \cdots \times S^{n_r}$ with antipodal action τ, and $Fl(n)$ with conjugation involution. Then $TC_{\mathbb{Z}_2}(S^{n_1} \times S^{n_2} \times \cdots \times S^{n_r} \times Fl(n)) = n(n-1) + r + k + 1$, where k is the number of even n_i's in $S^{n_1} \times S^{n_2} \times \cdots \times S^{n_r}$.*

Proof By Lemma 4.1 in [12], we have $TC_{\mathbb{Z}_2}(S^m) = 2$, if m is odd and $TC_{\mathbb{Z}_2}(S^m) = 3$, if m is even. Therefore, by Theorem 6, $TC_{\mathbb{Z}_2}(S^{n_1} \times S^{n_2} \times \cdots \times S^{n_r}) \leq r + k + 1$, where k is the number of even n_i's in $S^{n_1} \times S^{n_2} \times \cdots \times S^{n_r}$. By Proposition 6.3 in [4], $zcl_{\mathbb{Q}}(S^{n_1} \times S^{n_2} \times \cdots \times S^{n_r}) = r + k$, where k is the number of even n_i's in $S^{n_1} \times S^{n_2} \times \cdots \times S^{n_r}$. Therefore, the result follows from Theorem 10.

Acknowledgements The authors sincerely thank the anonymous referee for the valuable suggestions.

References

1. Bayeh M, Sarkar S (2015) Some aspects of equivariant LS-category. Topol Appl 196(part A):133–154
2. Colman H, Grant M (2012) Equivariant topological complexity. Algebraic Geom Topol 12(4):2299–2316
3. Cornea O, Lupton G, Oprea J, Tanré D (2003) Lusternik-Schnirelmann category. Math Surv Monogr AMS 103

4. Daundkar N, Sarkar S (2023) LS-category and topological complexity of several families of fibre bundles. arXiv:2302.00468
5. Davis DM (2010) Projective product spaces. J Topol 3(2):265–279
6. Dold A (1956) Erzeugende der Thomschen Algebra ℵ. Math Zeitschr 65:25–35
7. Fadell E (1985) The equivariant Ljusternik-Schnirelmann method for invariant functionals and relative cohomological index theories. Topological methods in nonlinear analysis Sm. Math Sup 95:41–70
8. Farber M (2003) Topological complexity of motion planning. Discrete Comput Geom 29(2):211–221
9. Fişekci S, Vandembroucq L (2021) On the LS-category and topological complexity of projective product spaces. J Homotopy Relat Struct 16(4):769–780
10. Fulton W (1997) Young tableaux with applications to representation theory and geometry. London Mathematical Society Student Texts 35, Cambridge University Press
11. González J, Grant M, Torres-Giese E, Xicoténcatl M (2013) Topological complexity of motion planning in projective product spaces. Algebraic Geom Topol 13(2):1027–1047
12. Lusternik L, Schnirelmann L (1934) Méthodes topologiques dans les problémes variationnels. Actualités scientifiques et industrielles, vol 188, Exposés sur l'analyse mathématique et ses applications, vol 3, Hermann, Paris, 42
13. Marzantowicz W (1989) A G-Lusternik-Schnirelman category of space with an action of a compact Lie group. Topology 28(4):403–412
14. Milnor JW, Stasheff JD (1974) Characteristic classes. Annals of Mathematics Studies, No. 76. Princeton University Press, Princeton, N. J.; University of Tokyo Press, Tokyo
15. Naskar B, Sarkar S (2020) On LS-category and topological complexity of some fiber bundles and Dold manifolds. Topology Appl 284:1–14
16. Nath A, Sankaran P (2019) On generalized Dold manifold. Osaka J Math 56:75–90
17. Sarkar S, Zvengrowski P (2022) On generalized projective product spaces and Dold manifolds. Homology Homotopy Appl 24(2):265–289

Real Dynamics of a Sixth-Order Family of Derivative-Free Iterative Method Without Memory

Sapan Kumar Nayak, P. K. Parida, Shwet Nisha, Chandni Kumari, Naveen Chandra Bhagat, and Babita Mehta

Abstract This manuscript introduces the dynamical behavior of a family of sixth-order derivative-free iterative method. When the proposed method is applied on a quadratic equation, and the presence of a parameter $\mu \in \mathbb{R}$, the iterative method creates the dynamical plane. Also, the reliability and stability of the iterative method have been studied using different tools. Moreover, information like convergence to n-cycles, different types of fixed points, and the chaotic nature of polynomials are all studied using the convergence plane.

Keywords Real dynamics · Non-linear equation · Iterative method · Convergence Plane · Bifurcation diagram · Lyapunov exponent · Basin of attraction

MSC: 65H05 · 26A33

1 Introduction

The most important problem in science and, basically, in mathematics is solving the non-linear equation

$$H(t) = 0, \tag{1}$$

where $H : \mathbb{R} \to \mathbb{R}$ be a continuous function. Often, to handle such types of non-linear equations, iterative methods are very useful, and one of the most common iterative method is the Newton method

S. Kumar Nayak · P. K. Parida (✉) · N. Chandra Bhagat · B. Mehta
Department of Mathematics, Central University of Jharkhand, Ranchi 835222, India
e-mail: pkparida@cuj.ac.in

S. Nisha
Department of Mathematics, Gaya College of Engineering, Gaya, India

C. Kumari
GLA University, Mathura, India

© The Author(s), under exclusive license to Springer Nature Singapore Pte Ltd. 2024
D. Giri et al. (eds.), *Proceedings of the Tenth International Conference on Mathematics and Computing*, Lecture Notes in Networks and Systems 964,
https://doi.org/10.1007/978-981-97-2066-8_4

$$t_{i+1} = t_i - \frac{H(t)}{H'(t)}. \tag{2}$$

Basically, the iterative methods transform Eq. (1) to a rational function of fixed point type. The dynamical characteristics of the rational function linked to an iterative process from a numerical perspective provide important insight into the stability and reliability of the method. And the dynamics of such types of rational functions are not well known because they are chaotic in nature. But there is a tool that gives us an opportunity to study the dynamical properties of such types of parameter-based rational operators. Let us suppose that $t_{i+1} = \Phi(t_i)$ be an iterative function, where $\Phi(t_i)$ is the fixed point function and t_{i+1} is determined by new information at $t_i, f_1(t_i), \ldots, f_n(t_i), n \geq 1$. Note that no previous data is reused. Thus,

$$t_{i+1} = \Phi(t_i, f_1(t_i), \ldots, f_n(t_i)).$$

Thus, Φ is called an iterative function without memory. Many mathematicians [1–6] have investigated the dynamics of iterative methods for solving non-linear equations in the complex plane. So the study of dynamical properties of non-linear equations using different types of iterative methods has been increased day-to-day, but the dynamical behavior of rational operator in real line is quite an interesting and challenging part. According to common belief, complex dynamics include real dynamics and both are the same. But the fact is not true. For example, in real dynamics, the point $z = \infty$ can be studied as another point in the complex plane, but it can't be done in the real line. Monotone convergence can be shown on a real line, but it doesn't exist in the complex plane. In real dynamics, there are also asymptotes, but in complex dynamics, the idea of an asymptote doesn't make sense. So, recently some authors have also devoted their research works to the real dynamics of different types of iterative methods ([3, 7–9], see references therein).

Parameter spaces have led to iterative algorithms in the complex plane whose dynamics are insufficiently understood. By assigning a complex parameter value to each point on the plane, we can investigate the orbits of the free critical points in these parameter spaces. Variations such as convergence to n-cycles, convergence to ∞, and even chaotic behavior have been discovered when mathematicians explore highly fascinating dynamical and parameter planes. In the real line, we have the opportunity to use tools like Feigenbaum diagrams and Lyapunov exponents [8] to better understand the behavior of one particular value of a parameter.

In this paper, we have studied the dynamical behavior of a three-step efficient family of iterative method [10] which needn't require any derivative evaluation

$$\begin{cases} y_i = t_i - \frac{H(t_i)}{H[t_i, u_i]}, \quad u_i = t_i - \mu H(t_i), \quad \mu \in \mathbb{R} - \{0\} \\ z_i = y_i - \frac{H(y_i)}{H[u_i, y_i]}, \\ t_{i+1} = z_i - \frac{H(z_i)}{H[u_i, z_i] + H[z_i, y_i] - H[u_i, y_i]}, \end{cases} \tag{3}$$

with sixth-order convergence, and take only four function evaluations during each iteration containing efficiency index=1.5651. The idea of Traub [11] is used to calculate the efficiency index of the iterative methods, using the formula $E_i = \theta^{1/d}$, where θ denotes the method's order of convergence and d denotes the total cost of function evaluations in each iteration.

1.1 Basic Definition and Results

A map $H : \mathbb{R} \to \mathbb{R}$ is called a **rational map** if $H(t) = \frac{P(t)}{Q(t)}$ where $P(t)$ and $Q(t) \neq 0$ are polynomials with complex coefficients and no common factors. Degree of $H(t)$ is given by $deg(H) = \max\{deg(P), deg(Q)\}$. For a given rational map $H : \mathbb{R} \to \mathbb{R}$, the **orbit** of a point $t \in \mathbb{R}$ is defined as $orb(t) = \{t, H(t), H^2(t), \ldots, H^n(t), \ldots\}$ where H^n denotes n-fold iterates of H. A point $t_0 \in H$ is known as a **fixed point** of H if $H(t_0) = t_0$. On the basis of associated multiplier $|H'(t)|$, a fixed point t_0 is called

- **Attracting** if $|H'(t_0)| < 1$;   **Super-attracting** if $|H'(t_0)| = 0$;
- **Repelling** if $|H'(t_0)| > 1$;   **Parabolic** if $|H'(t_0)| = 1$.

A point t_0 is called a **critical point** if the derivative of rational map vanishes at that point, i.e., $H'(t_0) = 0$. If t_0 is a point such that $H^n(t_0) = t_0$ but $H^k(t_0) \neq t_0$, for each of $k < n$, then t_0 is called a **periodic point** of period n. If $t_0 \in \mathbb{R}$ is a periodic point of period $n \geq 1$, then it is a fixed point of H^n. A periodic point of period n is *attracting, super-attracting, repelling, or parabolic* if it is as a fixed point of H^n is *attracting, super-attracting, repelling, or parabolic*. The orbit of a periodic point of period n is called **n-cycle**. The basin of attraction of an attractor β is the set of all initial conditions such that the method starting at t_0 converges to β. Mathematically *basin of attraction* for an attractor β is defined as $\mathcal{B}(\beta) = \{t_0 \in \mathbb{R} : H^n \to \beta, n \to \infty\}$. The **immediate basin of attraction** is the connected component of the basin of attraction which contains the attractor β. **Fatou set** of a rational map H is the set of point $t \in \mathbb{R}$ whose orbits tend to an attractor (fixed point, periodic point, or infinity) and denoted as $\mathcal{F}(H)$. A **Julia set** of a rational map H, denoted by $\mathcal{J}(H)$, is the closure of the set consisting of its repelling periodic points. Chaotic dynamics of H are contained in its Julia set.

   The rest of the paper is designed as follows: Sect. 2 devotes itself to the study of fixed points and critical points; in Sect. 3, we have represented the dynamical behavior of rational operator through the convergence plane; and Sect. 4 ends with the conclusion.

2 Study of Fixed and Critical Points

As indicated in the preceding section, we will investigate the behavior of the rational functions generated by the polynomials $q_0(t) = t^2$, $q_+(t) = t^2 + 1$, and $q_-(t) = t^2 - 1$ on the family of iterative method shown in (3). In the case of $q_0(t)$, $S_{q_0}(t, \mu) = \frac{t(\mu t - 1)}{2\mu t - 6}$ and the solution of the equation is $t = 0$, $t = \frac{1}{\mu}$ which is unstable for $\mu \to 0$. The roots of the polynomial $q_+(t)$ contain only complex roots and the corresponding rational operator

$$
S_{q_+}(t, \mu) = \frac{-\mu^4 + 6\mu^2 + \mu^4 t^{10} - 4\mu^3 t^9 + 3\mu^2 \left(\mu^2 + 2\right) t^8 - 4\mu t^7 + \left(2\mu^4 - 24\mu^2 + 1\right) t^6}{2\left(2\mu + \mu^2 t^5 - 2\mu t^4 + (2\mu^2 + 1) t^3 + (\mu^2 - 3) t\right)\left(\mu^2 + \mu^2 t^4 - 4\mu t^3 + (2\mu^2 + 3) t^2 - 4\mu t - 1\right)}
$$
$$
+ \frac{12\mu \left(2\mu^2 + 3\right) t^5 - \left(2\mu^4 + 60\mu^2 + 15\right) t^4 + 4\mu \left(8\mu^2 + 5\right) t^3 - 3\left(\mu^4 + 8\mu^2 - 5\right) t^2 + 4\mu \left(3\mu^2 - 5\right) t - 1}{2\left(2\mu + \mu^2 t^5 - 2\mu t^4 + (2\mu^2 + 1) t^3 + (\mu^2 - 3) t\right)\left(\mu^2 + \mu^2 t^4 - 4\mu t^3 + (2\mu^2 + 3) t^2 - 4\mu t - 1\right)}
$$

has four distinct complex strange fixed points for all values of $\mu \in \mathbb{R} - \{0\}$. Lastly, we will look at how the function $S_H(t, \mu)$ affects the polynomial $q_-(t)$. And the associated rational function is

$$
S_{q_-}(t, \mu) = \frac{\mu^4 + 6\mu^2 + \mu^4 t^{10} - 4\mu^3 t^9 - 3\mu^2 \left(\mu^2 - 2\right) t^8 - 4\mu t^7 + \left(2\mu^4 + 24\mu^2 + 1\right) t^6}{2\left(\mu^2 + \mu^2 t^4 - 4\mu t^3 + (3 - 2\mu^2) t^2 + 4\mu t + 1\right)\left(2\mu + \mu^2 t^5 - 2\mu t^4 + (1 - 2\mu^2) t^3 + (\mu^2 + 3) t\right)}
$$
$$
+ \frac{12\mu \left(2\mu^2 - 3\right) t^5 + \left(2\mu^4 - 60\mu^2 + 15\right) t^4 - 4\mu \left(8\mu^2 - 5\right) t^3 - 3\left(\mu^4 - 8\mu^2 - 5\right) t^2 + 4\mu \left(3\mu^2 + 5\right) t + 1}{2\left(\mu^2 + \mu^2 t^4 - 4\mu t^3 + (3 - 2\mu^2) t^2 + 4\mu t + 1\right)\left(2\mu + \mu^2 t^5 - 2\mu t^4 + (1 - 2\mu^2) t^3 + (\mu^2 + 3) t\right)}
$$

Since the roots of polynomial $q_-(t)$, $t = \pm 1$, the fixed points of the operator $S_{q_-}(t, \mu)$ are the roots of the $q_-(t)$.

In the Fig. 1, the behavior of the operator $S_{q_-}(t, \mu)$ is shown, where we can see the number strange fixed points depend on the parameter μ. Moreover, the real strange fixed points are

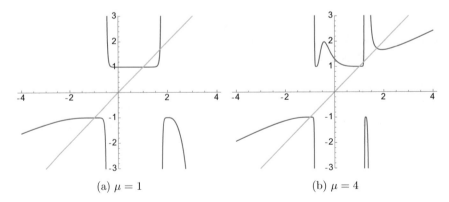

(a) $\mu = 1$ (b) $\mu = 4$

Fig. 1 For different values of parameter μ, $S_{q_-}(t, \mu)$ change the shape

$$sfp_1(\mu) = \frac{1}{\mu} - \frac{1}{2}\sqrt{\frac{4}{\mu^2} - \frac{2(3-2\mu^2)}{3\mu^2} + E_1} - \frac{1}{2}\sqrt{\frac{8}{\mu^2} - \frac{4(3-2\mu^2)}{3\mu^2} - E_1 - \frac{\frac{64}{\mu^3} - \frac{32}{\mu} - \frac{16(3-2\mu^2)}{\mu^3}}{4\sqrt{\frac{4}{\mu^2} - \frac{2(3-2\mu^2)}{3\mu^2}}E_1}}$$

$$sfp_2(\mu) = \frac{1}{\mu} - \frac{1}{2}\sqrt{\frac{4}{\mu^2} - \frac{2(3-2\mu^2)}{3\mu^2} + E_1} + \frac{1}{2}\sqrt{\frac{8}{\mu^2} - \frac{4(3-2\mu^2)}{3\mu^2} - E_1 - \frac{\frac{64}{\mu^3} - \frac{32}{\mu} - \frac{16(3-2\mu^2)}{\mu^3}}{4\sqrt{\frac{4}{\mu^2} - \frac{2(3-2\mu^2)}{3\mu^2}}E_1}}$$

$$sfp_3(\mu) = \frac{1}{\mu} + \frac{1}{2}\sqrt{\frac{4}{\mu^2} - \frac{2(3-2\mu^2)}{3\mu^2} + E_1} - \frac{1}{2}\sqrt{\frac{8}{\mu^2} - \frac{4(3-2\mu^2)}{3\mu^2} - E_1 + \frac{\frac{64}{\mu^3} - \frac{32}{\mu} - \frac{16(3-2\mu^2)}{\mu^3}}{4\sqrt{\frac{4}{\mu^2} - \frac{2(3-2\mu^2)}{3\mu^2}}E_1}}$$

$$sfp_4(\mu) = \frac{1}{\mu} + \frac{1}{2}\sqrt{\frac{4}{\mu^2} - \frac{2(3-2\mu^2)}{3\mu^2} + E_1} + \frac{1}{2}\sqrt{\frac{8}{\mu^2} - \frac{4(3-2\mu^2)}{3\mu^2} - E_1 + \frac{\frac{64}{\mu^3} - \frac{32}{\mu} - \frac{16(3-2\mu^2)}{\mu^3}}{4\sqrt{\frac{4}{\mu^2} - \frac{2(3-2\mu^2)}{3\mu^2}}E_1}}$$

where $D_1 = 64\mu^6 + 288\mu^4 + 270\mu^2 + 6\sqrt{3}\sqrt{64\mu^6 + 207\mu^4 + 27\mu^2 + 27}$

and $E_1 = \frac{9+48\mu^2+16\mu^4}{3\mu^2 D_1^{1/3}} + \frac{D_1^{1/3}}{3\mu^2}$.

Lemma 1 *The number of real simple strange fixed points of $S_{q_-}(t, \mu)$ is*

- *zero if $\mu \leq 0$,*
- *four strange fixed points $sfp_1(\mu)$, $sfp_2(\mu)$, $sfp_3(\mu)$, and $sfp_4(\mu)$, when $\mu > 0$.*

Next, to obtain the critical points of the operator $S_{q_-}(t, \mu)$ by its first-order derivative as follows:

$$S'_{q_-}(t, \mu) = \frac{\left(t^2 - 1\right)^5 \left(-\mu^2 + \mu^2 t^2 - 6\mu t + 3\right)\left(-\mu^2 + \mu^2 t^2 - 2\mu t + 1\right)^3}{2\left(\mu^2 + \mu^2 t^4 - 4\mu t^3 + (3 - 2\mu^2)t^2 + 4\mu t + 1\right)^2\left(2\mu + \mu^2 t^5 - 2\mu t^4 + (1 - 2\mu^2)t^3 + (\mu^2 + 3)t\right)^2}$$

Lemma 2 • *If $\mu \leq 0$ or $\mu = -1$, then there are no free critical points.*
• *If $\mu > 0$ or $\mu < -1$*

1. $cr_1 = \frac{1-\mu}{\mu}$,

2. $cr_2 = \frac{1+\mu}{\mu}$,

3. $c_3 = \frac{3\mu - \sqrt{\mu^4 + 6\mu^2}}{\mu^2}$,

4. $cr_4 = \frac{3\mu + \sqrt{\mu^4 + 6\mu^2}}{\mu^2}$.

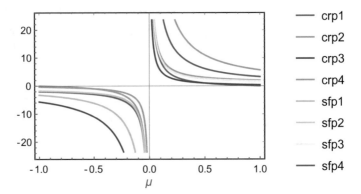

Fig. 2 Critical and Strange fixed points associated with the operator $S_{q_-}(t, \mu)$

The characteristics and connections between the genuinely strange fixed and free critical points are shown in Fig. 2 with different colors.

3 The Convergence Plane

Definition 1 Feigenbaum diagrams: Fiegenbaum diagram shown in Fig. 3, also known as bifurcation diagrams, are a powerful tool for investigating the real dynamics of iterative process families with a single parameter. The behavior of a point's orbits, including whether or not they cycle or are chaotic, can be seen graphically using this tool. This tool has two main drawbacks: it only displays one point per graphic, and if the graphic scale is too small, conflictive zones (convergence to anything other than a root) may be undetectable. Lyapunov exponents can solve the second issue (Figs. 4).

Definition 2 Lyapunov Exponents: It is commonly known that given an attracting (or repelling) cycle, the derivative will control the distance of the orbits of points in a neighborhood. In addition, let the beginning point's orbit $(t_0, t_1, ..., t_n - 1)$ is t_0. Then an orbit $t_0, t_1, ..., t_n - 1$ has a Lyapunov exponent defined as follows"

$$h(t_0) = log(L(t_0)) = lim \frac{1}{n}(log|S'(t_0)| + log|S'(t_1)| + log|S'(t_2)| + ... + log|S'(t_n)|)$$

Attracting orbits will have negative Lyapunov exponents and repelling orbits will have positive Lyapunov exponents. The main problem that presents this tool is that it only gives the information of one point at each graphic and the behavior of two near points could be very different.

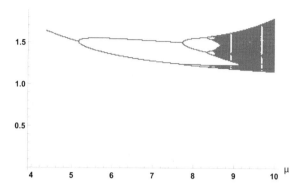

Fig. 3 Feigenbaum diagram of $\mathcal{S}_{q_-}(t, \mu)$

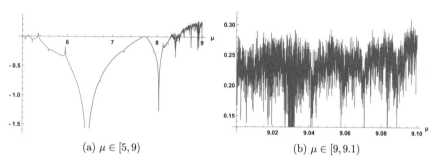

(a) $\mu \in [5, 9)$ (b) $\mu \in [9, 9.1)$

Fig. 4 Lyapunov exponents

3.1 Convergence Plane of Quadratic Polynomial for Distinct Real Roots

The convergence plane of the proposed method is plotted with a mesh of 600×600 real points using a Lenovo Ideapad flex 5, 1.19 GHz, Intel(R) Core(TM) i5-1035G1 CPU. The stopping criteria of the convergence plane are a tolerance of 10^{-3} and a maximum of 1000 iterations. The term "stable points" refers to the collection of all starting points for which the iterative process generates a convergence region. Initial guesses and the parameter μ both play a major role in determining whether iterative methods are stable. As a result of this, we have shown the stability and dynamical properties of the method by using the convergence planes. The convergence plane of the operator $\mathcal{S}_{q_-}(t, \mu)$ is painted with different colors:

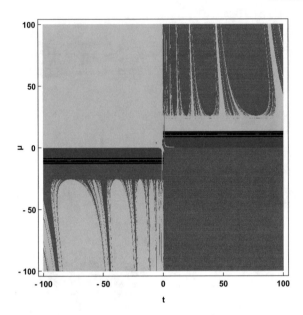

Fig. 5 Convergence plane of the $S_{q_-}(t, \mu)$ in the wide range

- The fixed points convergence to roots painted with individual color $t_1 = 1$ (*red*), $t_2 = -1$ (*green*).
- Blue, if the method convergence to 2-cycles, convergence to 3-cycles in cyan, convergence to 4-cycles in magenta, convergence to 5-cycles in orange, convergence to 6-cycles in dark blue, convergence to 7-cycles in dark green, and convergence to 8-cycles in dark yellow.
- The black color represents zones are the divergent of method.
- White color, for other cases.

In Fig. 5, we have illustrated the convergence plane of $S_{q_-}(t, \mu)$ on a larger scale of initial guesses and value of μ where symmetric behavior of method is found with a wide basin of attractions. In Fig. 6, the convergence plane is shown on a small scale, so that we can observe the basin of attraction of roots as well as different n-cycles of the operator $S_{q_-}(t, \mu)$.

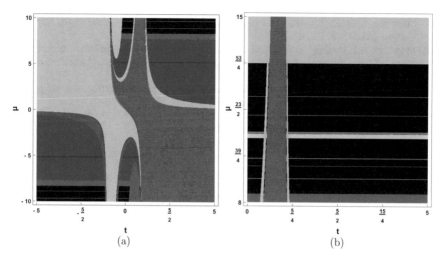

(a)　　　　　　　　　(b)

Fig. 6 Convergence plane of the $\mathcal{S}_{q_-}(t, \mu)$ with different values of μ

4　Conclusions

In this article, we have studied the dynamical behavior of a derivative-free higher order family of iterative method. The influence of the quadratic equation on the proposed method shows very interesting dynamical properties, i.e., the convergence of fixed and strange fixed points and method convergence to different n-cycles. The parameter of the proposed method that generates the number of fixed points is visualized with a graph. Moreover, the global convergence, stability, and reliability of the derivative-free method are presented with the help of the convergence plane.

References

1. Cordero A, Lucia G, Juan RT (2018) Choosing the most stable members of Kou's family of iterative methods. J Comput Appl Math 330:759–769
2. Babajee DKR, Cordero A, Torregrosa JR (2016) Study of iterative methods through the Cayley Quadratic Test. J Comput Appl Math 291:358–369
3. Maroju P, Magreñán ÁA, Sarría Í, Kumar A (2020) Local convergence of fourth and fifth order parametric family of iterative methods in Banach spaces. J Math Chem 58(3):1572–8897
4. Lee M, Kim YI (2020) The dynamical analysis of a uniparametric family of three-point optimal eighth-order multiple-root finders under the Möbius conjugacy map on the Riemann sphere. Numer Algorithms 83(3):1063–1090
5. Sharma D, Parhi SK, Sunanda SK (2020) A new class of fifth and sixth order root-finding methods with its dynamics and applications. Contemp Math, 401–416
6. Li J, Wang X, Madhu K (2019) Higher-order derivative-free iterative methods for solving nonlinear equations and their basins of attraction. Mathematics 7:1052
7. Magreñán ÁA, Gutiérrez JM (2015) Real dynamics for damped Newton's method applied to cubic polynomials. J Comput Appl Math 275:527–538

8. Magreñán ÁA (2014) A new tool to study real dynamics: the convergence plane. Appl Math Comput 248:215–224
9. Magreñán ÁA, Cordero A, Gutiérrez J, Torregrosa JR (2014) Real qualitative behavior of a fourth-order family of iterative methods by using the convergence plane. Math Comput Simul 105:49–61
10. Soleymani F (2011) Efficient sixth-order nonlinear equation solvers free from derivative. World Appl Sci J 13(12):2503–2508
11. Jain MK, Iyengar SRK, Jain RK (2012) Numerical methods: for scientific and engineering computation, 6th edn. New Age International Publishers, New Delhi

Algorithm for Reconstruction Number of Split Graphs

V. Manikandan and S. Monikandan

Abstract A *card* $G - v$ of a graph G is obtained by deleting the vertex v and all edges incident with v. The multiset of all cards of G is called the *deck* of G. A graph is *reconstructible* if it is determined up to isomorphism from the collection of all its cards. The *Reconstruction Conjecture* asserts that all graphs of order at least three are reconstructible. The minimum number of cards of G that do not belong to the deck of any graph not isomorphic to G is called the *reconstruction number* of G. A *split graph* is a graph in which the vertices can be partitioned into an independent set and a clique. In this paper, we prove that the degree sequence of a split graph G can be found by using some six cards of G. We give an algorithm to find the reconstruction number of split graphs G which uses only six cards of G for most of the cases.

Keywords Split graph · Reconstruction number · Clique · Independent set

MSC Classification 05C78 · 05C75 · 05C60

1 Introduction

All graphs considered in this paper are finite, simple and undirected. Terms not defined here are taken as in [13]. By *deg* v, we mean the degree of a vertex v in a graph G. The *degree sequence* of a graph G is denoted by $DS(G)$ and represented by $(d_1, d_2, d_3, \ldots, d_n)$, where $d_i \geq d_{i+1}$ and n is the order of G. The set of all vertices adjacent to v in G is denoted by $N_G(v)$ and it is called the *neighbourhood* of v in G. A *clique* of a graph G is a vertex subset inducing a complete subgraph of G. The maximum cardinality of a clique in G is called the *clique number* of G and it is denoted by $\omega(G)$. A subset I of $V(G)$ is called an *independent set* if no pair of distinct vertices of I are adjacent in G.

V. Manikandan (✉) · S. Monikandan
Department of Mathematics, Manonmaniam Sundaranar University, Tirunelveli 627012, Tamil Nadu, India
e-mail: manikandanv1935@gmail.com

© The Author(s), under exclusive license to Springer Nature Singapore Pte Ltd. 2024   45
D. Giri et al. (eds.), *Proceedings of the Tenth International Conference on Mathematics and Computing*, Lecture Notes in Networks and Systems 964,
https://doi.org/10.1007/978-981-97-2066-8_5

A *card* $G - v$ of a graph G is obtained from G by deleting a vertex v and all edges incident with v. The multiset of all cards of a graph G, denoted by $\mathscr{D}(G)$, is called the *deck* of G. If G and H are graphs with $\mathscr{D}(H) = \mathscr{D}(G)$, then G is a *reconstruction* of H and vice-versa. The collection of all reconstructions of a graph G is denoted by $Rec(G)$. A graph G is *reconstructible* whenever H is a graph with $\mathscr{D}(H) = \mathscr{D}(G)$ implies $H \cong G$. The two graphs K_2 and $2K_1$ are not isomorphic but their decks agree and hence they are not reconstructible. Let $\mathscr{F}$ be a family of graphs. Then $\mathscr{F}$ is *recognizable* if, for each $F \in \mathscr{F}$, the graph E is in $\mathscr{F}$ for every $E \in Rec(F)$. A family $\mathscr{F}$ is *weakly reconstructible* if, for each $F \in \mathscr{F}$ and for each $E \in Rec(F)$ with $E \in \mathscr{F}$, we have $E \cong F$. A family $\mathscr{F}$ is *reconstructible* if and only if it is both recognizable and weakly reconstructible.

The famous *Reconstruction Conjecture* (RC) [7] asserts that every graph of order more than two is reconstructible. The manuscripts [3, 9] are surveys of work done on the RC and related problems. Recently, Devi Priya and Monikandan gave a reduction of the RC in [5].

For a reconstructible graph G, the minimum number of cards of G that do not belong to the deck of any graph not isomorphic to G is called the *reconstruction number* of G, denoted by $rn(G)$. This parameter of G was introduced by Harary and Plantholt [8]. Bollobas [2] proved that almost all graphs have reconstruction number 3. Recently, Monikandan and Anu [10] have proved that $rn(G) = 3$ for all connected graphs G with at least three pendant vertices if $rn(H) = 3$ for all 2-connected graphs H and for some graphs H with exactly two blocks, one of them is K_2. This results strengthen the above results of Bollobas [2] as well as Myrvold [12]. For a survey of early results on the reconstruction number problems, see [1, 9, 11]. In this paper, we prove that split graphs G are recognizable by any set of six cards of G and that the clique number of a split graph G can be found by using some six cards of G. We also prove that $rn(G) = 3$ for all graphs G with $\omega(G) \geq n - 1$. We also prove that the degree sequence of a split graph G can be found by using some six cards of G. Finally, we give an algorithm to find the reconstruction number of split graphs G which uses minimum six cards of G.

2   Graph with Clique Number at Least $n - 1$

As a prelude, we prove that all graphs G with $\omega(G) \geq n - 1$ have $rn(G) = 3$.

Theorem 1 *For graphs G with $\omega(G) \geq n - 1$, $rn(G) = 3$.*

Proof A graph of order n has a complete card if and only if $\omega(G) \geq n - 1$.

Moreover, at least three cards of G are complete if and only if $\omega(G) = n$.
Thus, whether $\omega(G) = n$ or $n - 1$ can be recognized from some set of three cards of G.

If $\omega(G) = n$, then G is complete and it is known [4] that $rn(K_n) = 3$.
So, let us take that $\omega(G) = n - 1$.
We shall prove by three cases as below.

Case 1. $G \cong K_{n-1} \cup K_1$.

The cards selection: $G - v_s$, $G - v_t$, $G - v_l$, where v_s, v_t are vertices of degree $n - 2$ and v_l is the isolated vertex.

Now the card $G - v_l$ is complete and so G can have at most one isolated vertex. The unique isolated vertex in the card $G - v_i$ would not be adjacent with any vertex other than v_i in G, where $i \in \{s, t\}$. Consequently, the unique isolated vertex in these two cards would be the same vertex in G and it is the only isolated vertex of G. Thus G can be obtained uniquely from $G - v_l$ by adding a new vertex, which implies $rn(G) = 3$.

Case 2. $G \cong K_n - e$.

The cards selection: $G - v_s$, $G - v_t$, $G - v_l$, where v_s, v_t are the two vertices of degree $n - 2$ and v_l is any other vertex.

Since both $G - v_s$ and $G - v_t$ are complete, every vertex other than v_s, v_t would be a complete vertex in G. Thus, G can be obtained uniquely from $G - v_l$ by adding a new vertex and joining it to all the vertices. Hence $rn(G) = 3$.

Case 3. $G \not\cong K_{n-1} \cup K_1$ and $G \not\cong K_n - e$.

Now, since $\omega(G) = n - 1$, there exists a vertex v of degree k, where $0 < k < n - 2$ such that $G - v \cong K_{n-1}$.

The cards selection: $G - v_s$, $G - v_t$, $G - v_l$, where $deg_G v_s = n - 1$, $deg_G v_t = n - 2$ and $G - v_l$ is complete.

Existence of these three cards force that $G \not\cong K_{n-1} \cup K_1$ and $G \not\cong K_n - e$, but $G - v_l \cong K_{n-1}$. Now G can be obtained uniquely from $G - v_s$ by adding a new vertex and joining it to all the vertices in $G - v_s$. Hence $rn(G) = 3$. □

3 Recognition of Split Graphs from Any Six Cards

The following theorem was proved in [6].

Theorem 2 *A graph is split if and only if it has no induced subgraph isomorphic to C_5, C_4 or $2K_2$.*

Lemma 3 *Any nonsplit graph G can have at most five split cards.*

Proof By Theorem 2, the graph G has an induced subgraph H isomorphic to C_5, C_4 or $2K_2$. If $H \cong C_5$, then a card obtained from G by deleting a vertex from a copy of H in G may be possibly containing no induced subgraphs isomorphic to H. Consequently, any nonsplit graph can have at most five vertices such that the corresponding five cards may not contain H as an induced subgraph. Since the other two have four vertices of the same type, the graph G has at most five split cards. □

Corollary 4 *Split graphs are recognizable by at most six cards.*

Proof Consider any six cards of split graph G. If any one of these six cards is nonsplit, then G is nonsplit since every card of a split graph is split. Otherwise, G is split by Lemma 3. □

4 Clique Number from Some at Most Six Cards

In view of Theorem 1, we can assume that all split graphs G considered hereafter have clique number at most $n - 2$; let C be a maximal clique, let $I = V - C$ be an independent set of G and let $\omega(G) = k$. Throughout this paper, we use the notation G, C, I and k in the above sense.

It is clear that the clique number of any card of a split graph G is either $\omega(G)$ or $\omega(G) - 1$. In a split graph G, $d_i \geq \omega - 1$, for $i = 1, 2, 3, \ldots, \omega(G)$ and $d_i \leq \omega - 1$, for $i = \omega + 1, \omega + 2, \omega + 3, \ldots, n$. For a vertex $u \in C$, a vertex $v \in V - C$ is said to be an *sn-vertex* of u if $v \nsim u$ and $N_C(u) = N_C(v)$. In particular, for any card $G - x$,

$$\omega(G - x) = \begin{cases} \omega(G) - 1 & \text{if } x \in C \ \& \ x \ \text{has no } sn - \text{vertex} \\ \omega(G) & \text{otherwise.} \end{cases}$$

For every vertex $v \in I$, $\omega(G - v) = \omega(G)$ since C is maximal. Also, if every vertex in C has an *sn*-vertex, then $\omega(G - v) = \omega(G)$, $\forall v \in V(G)$. Therefore every card has equal clique number in this case. Otherwise, $\omega(G - v) = \omega(G) - 1$ for some vertex v in C, which prove the next lemma.

Lemma 5 *If there is a vertex $v \in C$ having no sn-vertex in G, then G has two cards with different clique number.*

Corollary 6 *If $\omega(G) > \frac{n}{2}$, then G has two cards with different clique number.*

Proof If $\omega(G) > \frac{n}{2}$, then not every vertex in C has an *sn*-vertex and the proof follows.   □

Theorem 7 *If every vertex in C has an sn-vertex, then the clique number of G is obtained by some at most six cards.*

Proof The proof uses some at most four cards but split graphs are recognizable by any set of at most six cards (Corollary 4), and so we have such assertion in the theorem. If any three cards are totally disconnected, then so is G and hence $\omega(G) = 1$. Any graph with a triangle can have at most three triangle-free cards. It follows that, if G has at least four triangle-free cards of which at least one has an edge, then G is triangle-free and hence $\omega(G) = 2$.

The cards selection: $G - v_r$, $G - v_s$, $G - v_t$, where $v_r \in C$ and let $v_s, v_t \in I$.

Since, in this case, all cards have the equal clique number, we have $\omega(G - v_r) = \omega(G - v_s) = \omega(G - v_t) = k$ (say). Then $\omega(G) \geq k$. Suppose, to the contrary, that $\omega(G) = k + 1$. Then, $d_j \geq k$ for $j = 1, 2, 3, \ldots, k + 1$ and $d_i \leq k$ for $i = k + 2, k + 3, k + 4, \ldots, n$. In $DS(G - v_r)$, kth entry is $k - 1$ and hence $d_{k+1} = k$ in G.

In each of the k cards $G - v_i$, for $i = 1, 2, \ldots k$, the entry in the kth place of the degree sequence is $k - 1$. In each of the $n - k - 1$ cards $G - v_i$, for $i = k + 2, k +$

3, ..., n, the entry in the $(k+1)$th place of the degree sequence is k.   ...(∗)
The entries in the degree sequence of the only remaining card $G - v_{k+1}$ are not
certain. But the degree sequences of the two cards $G - v_s$ and $G - v_t$ do not satisfy
none of the degree conditions in (∗), which is a contradiction and completing the
proof.   □

5   Degree Sequence from Some Six Split Cards

Label the vertices of G by $v_1, v_2, v_3, \ldots, v_n$ such that $d_1 \geq d_2 \geq d_3 \geq \cdots \geq d_n$,
where $d_i = deg\ v_i$. Consider a card $G - v_t$ and label its vertices by $v'_1, v'_2, v'_3, \ldots,$
v'_{n-1} such that $d'_1 \geq d'_2 \geq d'_3 \geq \cdots \geq d'_{n-1}$, where $d'_i = deg\ v'_i$.

Observation 8 *If v_t lies in C of G, then*

$$d'_i = \begin{cases} d_i - 1 & if\ i < t \\ d_{i+1} - 1 & if\ t \leq i \leq \omega - 1 \\ d_{i+1}\ or\ d_{i+1} - 1 & if\ \omega \leq i \leq n - 1 . \end{cases}$$

If v_t does not lie in C of G, then

$$d'_i = \begin{cases} d_i\ or\ d_i - 1 & if\ i \leq \omega \\ d_i & if\ \omega + 1 \leq i < t \\ d_{i+1} & if\ t \leq i \leq n . \end{cases}$$

Theorem 9 *The degree sequence of a split graph can be obtained from some six cards.*

Proof The proof uses some at most three cards but split graphs are recognizable by
any set of at most six cards (Corollary 4), and so we have such assertion.

Let the degree sequence of the vertices in I be $(n_1{}^{k_1}, n_2{}^{k_2}, n_3{}^{k_3}, \ldots, n_l{}^{k_l})$, where
$n_1 \geq n_2 \geq \cdots \geq n_l$ and $n_i{}^{k_i}$ means that there are k_i vertices of degree n_i and represents $\underbrace{n_i, n_i, n_i, \ldots, n_i}_{k_i\text{-times}}$. Then the degree sequence of G would be

$(d_1, d_2, \ldots, d_\omega, n_1{}^{k_1}, n_2{}^{k_2}, n_3{}^{k_3}, \ldots, n_l{}^{k_l})$, where $d_1 \geq d_2 \geq d_3 \geq \cdots \geq d_\omega$.
Therefore, for each $v \in I$, the degree sequence of $G - v$ is of the form
$(d'_1, d'_2, \ldots, d'_\omega, n_1{}^{k_1}, n_2{}^{k_2}, \ldots, n_h{}^{k_h-1}, \ldots, n_l{}^{k_l})$, where $d'_1 \geq d'_2 \geq d'_3 \geq \cdots \geq d'_\omega$.
    Consider the six cards $G - v_i$, where $i = 1, 2, \ldots, 6$, $v_r \in C$ and $v_\alpha, v_\beta \in I$.
Then, from the degree sequences of these six cards and observation 8, we can identify
whether $v_i \in C$ or $v_i \in I$ for each $i = 1, 2, \ldots, 6$.
Case 1. Not all the vertices in I have the same degree.

Now choose the vertices v_α, v_β in I such that $DS(G - v_\alpha) = (a_1, a_2, a_3, \ldots,$
$a_\omega, n_1{}^{k_1}, n_2{}^{k_2}, \ n_3{}^{k_3}, \ldots, n_s{}^{k_s-1}, \ldots, n_l{}^{k_l})$ and $DS(G - v_\beta) = (b_1, b_2, b_3, \ldots,$
$b_\omega, n_1{}^{k_1}, n_2{}^{k_2}, n_3{}^{k_3}, \ldots, n_t{}^{k_t-1}, \ldots, n_l{}^{k_l})$, where $s \neq t$. By comparing these two
degree sequences, we get the degrees of vertices in I of G.

Case 2. All the vertices in I have equal degree.

Now $DS(G - v_\alpha) = (c_1, c_2, c_3, \ldots, c_\omega, n_1{}^{n-\omega-1})$ and $DS(G - v_\beta) = (e_1, e_2, e_3,$
$\ldots, e_\omega, n_1{}^{n-\omega-1})$. By comparing these two degree sequences, we get each vertex
in I of G has degree n_1.

Thus, in both cases, we found the values of $d_{\omega+1}, d_{\omega+2}, d_{\omega+3}, \ldots, d_n$ in $DS(G)$.

Now consider the card $G - v_r$ where $v_r \in C$ and v_r has no sn-vertex (if any).
Let $DS(G - v_r) = (q_1, q_2, q_3, \ldots, q_{\omega-1}, q_\omega, q_{\omega+1}, q_{\omega+2}, \ldots, q_{n-1})$, where

$$q_i = \begin{cases} d_i - 1 & \text{if } i < r \\ d_{i+1} - 1 & \text{if } r \leq i \leq \omega - 1. \end{cases}$$

Now by comparing the entries of $DS(G - v_r)$ from the (ω)th place onwards with
the known sequence $(d_{\omega+1}, d_{\omega+2}, d_{\omega+3}, \ldots, d_n)$, we will get information about the
number of vertices (say s_1) in I to which v_r is adjacent. Hence $deg_G v_r = s_1 + \omega - 1$.
Now $DS(G)$ can be obtained by putting $deg\ v_r$ in the appropriate place in the non-
increasing sequence $(q_1 + 1, q_2 + 1, q_3 + 1, \ldots, q_{\omega-1} + 1, d_{\omega+1}, d_{\omega+2}, d_{\omega+3}, \ldots,$
$d_n)$. Thus, we have obtained $DS(G)$ by three cards $G - v_\alpha, G - v_\beta$ and
$G - v_r$. $\square$

6   An Algorithm to Find the Reconstruction Number of Split Graphs

For a graph F, the *extensions* of F, denoted by $Ex(F)$, is the set of non-isomorphic
graphs that results from adding one vertex to the graph F, and adding edges incident
to the new vertex in every possible way. For a card $G - x_i$, let A_i be the subset of
$Ex(G - x_i)$ such that $DS(H) = DS(G)$ for every $H \in A_i$. Clearly, $A_i \neq \emptyset$ as at
least G is in A_i. If $|A_i| = 1$ for some vertex x_i in G, then $rn(G) \leq 6$.

Algorithm:

Step 1. Choose three vertices x_1, x_2, x_3, where $x_1, x_2 \in I$ and $x_3 \in C$ such that
$deg(x_1) \neq deg(x_2)$ if possible as in Theorem 9 and also choose three more vertices $x_{t_1}, x_{t_2}, x_{t_3}$. The degree sequence of G is known from these six cards by Theorem 9. If there exist any such six vertices $x_1, x_2, x_3, x_{t_1}, x_{t_2}$ and x_{t_3} satisfying the
condition $|A_i \cap A_j \cap A_r \cap A_{t_1} \cap A_{t_2} \cap A_{t_3}| = 1$, then $(A_i \cap A_j \cap A_r \cap A_{t_1} \cap A_{t_2} \cap$
$A_{t_3}) = \{G\}$ and hence $rn(G) \leq 6$. Otherwise $rn(G) > 3$ and move to the next step.

Step 2. Choose a vertex $x_{\beta_1} \in V(G) - \{x_1, x_2, x_3, x_{t_1}, x_{t_2}, x_{t_3}\}$. If there exist any
such seven vertices $x_1, x_2, x_3, x_{t_1}, x_{t_2}, x_{t_3}, x_{\beta_1}$ satisfying the condition $|A_i \cap A_j \cap$
$A_r \cap A_{t_1} \cap A_{t_2} \cap A_{t_3} \cap A_{\beta_1}| = 1$, then G has $rn(G) \leq 7$. Otherwise, $rn(G) > 4$ and
move to the next step.

Step 3. Choose a vertex $x_{\beta_2} \in V(G) - \{x_1, x_2, x_3, x_{t_1}, x_{t_2}, x_{t_3}, x_{\beta_1}\}$. If there exist any such eight vertices $x_1, x_2, x_3, x_{t_1}, x_{t_2}, x_{t_3}, x_{\beta_1}, x_{\beta_2}$ satisfying the condition $|A_i \cap A_j \cap A_r \cap A_{t_1} \cap A_{t_2} \cap A_{t_3} \cap A_{\beta_1} \cap A_{\beta_2}| = 1$, then $rn(G) \leq 8$. Otherwise $rn(G) > 5$. Then we move to the next step and continue like this.

Step l: Choose a vertex $x_{\beta_{l-1}} \in V(G) - \{x_1, x_2, x_3, x_{t_1}, x_{t_2}, x_{t_3}, x_{\beta_1}, x_{\beta_2}, \dots, x_{\beta_{l-2}}\}$. If there exist any such $l + 5$ vertices $x_1, x_2, x_3, x_{t_1}, x_{t_2}, x_{t_3}, x_{\beta_1}, x_{\beta_2}, \dots, x_{\beta_{1l-1}}$ satisfying the condition $|A_i \cap A_j \cap A_r \cap A_{t_1} \cap A_{t_2} \cap A_{t_3} \cap A_{\beta_1} \cap A_{\beta_2} \cap \dots \cap A_{\beta_{l-1}}| = 1$, then G has $l + 2 \leq rn(G) \leq l + 5$. Otherwise $rn(G) > l + 2$. Then we move to the next step and continue like this.

Step $n - 5$. If G is reconstructible by the set $\{A_\alpha : x_\alpha \in V(G)\}$, then $n - 3 \leq rn(G) \leq n$. Otherwise, G is not reconstructible. $\square$

Remark 1 If $|A_{\alpha_1} \cap A_{\alpha_2} \cap \dots \cap A_{\alpha_k}| = 1$ for some vertex subset $S = \{x_{\alpha_1}, x_{\alpha_2}, \dots, x_{\alpha_k}\}$ of split graph G, then $rn(G) \leq k + 2$. If $rn(G) = h$, then the above algorithm must end at t steps, where $h - 5 \leq t \leq h - 2$ for $h \geq 6$.

7 Conclusion

One of the foremost unsolved problems in graph theory is the graph reconstruction conjecture, which was first proposed by P. J. Kelly and S. M. Ulam in 1941. It states that every graph with at least 3 vertices is reconstructible. Several classes of graphs are already proved to be reconstructible with the hope that one day enough classes to include, all graphs would come into the fold. But the reconstruction of general split graphs is still an open problem. In this paper, we have an algorithm for the reconstruction number of split graphs which uses minimum of six cards. A few classes of split graphs are already proved to be reconstructible. Using this algorithm, we can find the nearest bound for the reconstruction number of a split graph if it is reconstructible.

Declarations

Author Contributions

All authors have equal contributions. All authors read and approved the final manuscript.

Data Availability

The authors have not used any data for the preparation of this manuscript.

Compliance with Ethical Standards

Funding

This study was funded by National Board for Higher Mathematics(NBHM), Government of India (grant number: 02011/14/2022/NBHM(R.P)/R&D II).

Conflict of Interest

S. Monikandan has received research grants from National Board for Higher Mathematics(NBHM), Government of India. V. Manikandan declares that he has no conflict of interest.

Ethical Approval

This article does not contain any studies with human participants or animals performed by any of the authors.

References

1. Asciak KJ, Francalanza MA, Lauri J, Myrvold W (2010) A survey of some open questions in reconstruction numbers. Ars Combin 97:443–456
2. Bollobas B (1990) Almost every graph has reconstruction number three. J Graph Theory 14(1):1–4
3. Bondy JA (1991) A graph reconstructor's manual. In: Surveys in combinatorics (Proceedings of British combinatorial conference). London mathematical society lecture notes, no 116, pp 221–252
4. Bondy JA, Hemminger RL (1977) Graph reconstruction-a survey. J Graph Theory 1:227–268
5. Devi Priya P, Monikandan S (2018) Reconstruction of distance hereditary 2-connected graphs. Discret Math 341:2326–2331
6. Foldes S, Hammer PL (1976) Split graphs. University of Waterloo, CORR: 76-3
7. Harary F (1964) On the reconstruction of a graph from a collection of subgraphs. In: Fiedler M (ed) Theory of graphs and its applications. Academic Press, New York, pp 47–52
8. Harary F, Plantholt M (1985) The graph reconstruction number. J Graph Theory 9:451–454
9. Lauri J (1993) Vertex deleted and edge deleted subgraphs. Collected papers published on the occasion of the quatercentenary celebrations. University of Malta, pp 495–524
10. Monikandan S, Anu A (2021) Reconstruction Number of connected graphs with unique pendant vertex. Discret Appl Math 305:357–365. https://doi.org/10.1016/j.dam.2020.06.005
11. Myrvold WJ (1988) A report on the ally reconstruction problem. In: Alavi Y, Chartrand G, Oellermann OR, Schwenk AJ (eds) Graph theory, combinatorics and application, vol 2, pp 947–956
12. Myrvold WJ (1988) The ally and adversary reconstruction problems, Ph.D. thesis, University of Waterloo
13. West DB (2005) Introduction to graph theory, 2nd ed. Prentice-Hall

Computation of Captive, Half Certified and Majority Domination Numbers of a Family of 3-Regular Graphs

T. Kalaiselvi and Yegnanarayanan Venkatraman

Abstract Some interesting results about the computation of the captive domination number, strong captive domination number, half certified captive domination number and majority domination number are reported here for flower snark graphs that are a pertinent family of 3-regular graphs.

Keywords Captive domination number · Strong captive domination number · Half certified captive domination number · Majority domination number · Flower snarks

1 Introduction

Graph theory is a pertinent branch of mathematics and computer science. In the last few decades, we have seen a phenomenal growth. It is because of its application to tasks in optimization, combinatorial/algebraic challenges that are classic. One can see its applications in varied fields such as physical/engineering/social/biological sciences/linguistics, to name a few. Lately, the theory of domination has become the central research activity. In [5], the domination and irredundant number of 4-regular graphs were discussed. Given a graph $G = (V, E)$, a subset $S \subseteq V$ is termed as a dominating set if every vertex in V is in S or adjacent to some vertex in S. A dominating set of least size is termed a γ-set and the number of elements of any γ-set is called a domination number denoted by $\gamma(G)$. The task of finding a dominating set is in general a hard task. A dominating set $S \subseteq V(G)$ is called a total dominating set if any vertex $v \in V(G)$ has at least one adjacent element in S. The size of a minimal total dominating set is referred to as total domination number of G and it is denoted by $\gamma_t(G)$. A total dominating set S is called a captive dominating set if any vertex in S is

T. Kalaiselvi · Y. Venkatraman (✉)
Department of Mathematics, Kalasalingam Academy of Research and Education, Krishnankoil 626126, Tamilnadu, India
e-mail: prof.yegna@gmail.com

T. Kalaiselvi
e-mail: t.kalaiselvi@klu.ac.in

adjacent with at least one element in $V - S$. The size of a minimal captive dominating set is referred to as the captive domination number of G and denoted by $\gamma_{ca}(G)$. Suppose that $uv \in E(G)$ of G. u is said to dominate v strongly if $deg(u) \geq deg(v)$. Clearly any vertex of $V(G)$ dominates strongly itself. S is called a strong dominating set if each $v \in V - S$ is strongly dominated by some u in S. The strong domination number $\gamma_{std}(G)$ of G is the minimum size of a strong captive dominating set. S is termed a strong captive dominating set if it is both strong and a captive dominating set. The least number of elements in such an S is called strong captive domination number, $\gamma_{sca}(G)$. A captive dominating set S is called a half certified captive dominating set if S is a captive dominating set and every vertex in S has at least two neighbors in $V - S$. The size of a minimal half certified captive dominating set is called the half certified captive domination number of G and denoted by $\gamma_{hcca}(G)$. The neighborhood of v is the set $N_G(v) = N(v) = \{u \in V(G) : uv \in E(G)\}$. If $S \subseteq V(G)$, then the open neighborhood of S is the set $N_G(S) = N(S) = \bigcup_{v \in S} N_G(v)$. The closed neighborhood of S is $N_G[S] = N[S] = S \cup N(S)$. A subset S of $V(G)$ is a majority dominating set if at least half of the vertices of $V(G)$ either belong to S or are adjacent to the elements of S. That is $|N[S]| \geq \left\lceil \dfrac{V(G)}{2} \right\rceil$. The minimum size of cardinality of a majority dominating set of G is called the majority domination number of G and is denoted by $\gamma_m(G)$.

The study of regular graphs and their properties with respect to the concept of domination is available in plenty in the literature. For instance, the domination number in 4-regular Knodel graphs $W_{4,n}$ is found in [6, 11, 12]. Knodel graphs, $W_{\Delta,n}$, are one of the three pertinent families of graphs that possess wonderful attributes in terms of gossiping and broadcasting. One can also see [10, 16] for more. Like Knodel graphs, Flower Snark graphs are also one of the interesting families of graphs, and some investigation about this with respect to certain domination parameters is reported here. A few other interesting investigations about regular graphs can be found in [1, 8, 14].

The idea of captive dominating set was first introduced in [2]. In [15] the authors computed captive domination number for certain classes of graphs such as paths, cycles, n-dimensional cubes, product graphs and for graphs that result out of special operations like fan graph, helm graph, friendship graph, book graph, flower graph, windmill graph, etc. The concept of strong captive domination number of graph was introduced by [15]. In [15], the authors computed it for various classes of graphs such as cycles, paths, special subgraph of a complete equibipartite graph, double star graph, comb graph, generalized caterpillar graph, binomial trees, Mobius ladder graph and platonic graphs. The idea of half certified captive dominating set was first introduced in [7]. In [7], the authors discussed the existence or non-existence of half certified captive domination number for certain classes of graphs and computed half certified captive domination number for several classes of graphs.

2 Flower Snark Graphs

Isaacs in the year 1975 announced the flower snark graphs as a notable instance of a cubic, no cut-edge family of graphs, that admits no 3-edge-coloring. It was formally denoted as J_n, on $4n$ vertices. Here, n is odd and $n \geq 5$. The vertices of J_n are labeled as $V_i = \{z^i, w^i, x^i, y^i\}$ for $1 \leq i \leq n$. Its edges are partitioned as n star graphs and two cycles as explained below. Each 4-tuple V_i of $V(J_n)$ induces a $K_{1,3}$ with x^i as its center. The vertices y^i induce an odd cycle $C = y^1 y^2 \ldots y^n y^1$. The vertices z^i and w^i induce an even cycle $C_{2n} = z^1 z^2, \ldots z^n w^1 w^2 \ldots w^n z^1$. Figure 1 shows a flower snark J_3 on 12 vertices drawn by Isaacs. It depicts the flower shape and the crossing of edges (z^1, w^3) and (w^1, z^3). Figure 2 shows another way of drawing flower snarks as a sequential string of stars defined by $E(J_n) = \{y^i y^{i+1}$ for $1 \leq i \leq n - 1, y^n y^1; z^i z^{i+1}$ for $1 \leq i \leq n - 1, z^n w^1; w^i w^{i+1}$ for $1 \leq i \leq n - 1, w^n z^1; x^i y^i, x^i z^i, x^i w^i$ for $1 \leq i \leq n\}$. In [13], all flower snarks that are 4-flow-critical are discussed. In [4] the pebbling number of an n-dimensional Flower Snark J_n for $n \geq 5$ was found. In [9], the various Roman domination numbers were computed for flower snark graphs.

Fig. 1 The graph J_3

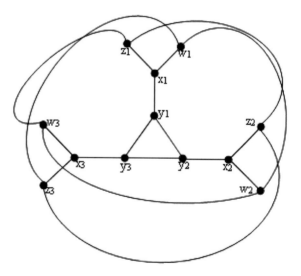

Fig. 2 The flower snark
graph J_3

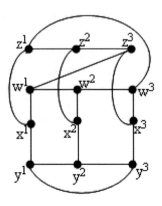

3   Results and Proofs

Theorem 1 *If J_n is a flower snark graph then*

$$
\gamma_{ca}(J_n) = \begin{cases}
8k + 2 & \text{if } n \equiv 1 \ (mod \ 4) \\
8k + 4 & \text{if } n \equiv 2 \ (mod \ 4) \\
6k & \text{if } n \equiv 0 \ (mod \ 4) \\
6k + 5 & \text{if } n \equiv 3 \ (mod \ 4)
\end{cases}
$$

Proof Consider for $n \geq 3$, the graph $n K_{1,3}$. Let x^i stand for the three-degree vertex in the ith copy of $K_{1,3}$. Also let y^i, z^i, w^i stand for the remaining three vertices in the ith copy of $K_{1,3}$. Hence $V(J_n) = \{z^i, w^i x^i, y^i \text{ for } 1 \leq i \leq n\}$. So $|V(J_n)| = n + 3 + 3 + \cdots + 3 \, (n \text{ times}) = n + 3n = 4n$. Let $E(J_n) = \{y^i y^{i+1} \text{ for } 1 \leq i \leq n - 1, \, y^n y^1; \, z^i z^{i+1} \text{ for } 1 \leq i \leq n - 1, \, z^n w^1; \, w^i w^{i+1} \text{ for } 1 \leq i \leq n - 1, \, w^n z^1; \, x^i y^i, x^i z^i, x^i w^i \text{ for } 1 \leq i \leq n\}$. Then $|E(J_n)| = 6n$. The graph J_3 is shown in Fig. 2.

In order to construct a dominating set S for J_n, it is essential to include at least one vertex from each copy of $K_{1,3}$ in J_n. For this we select the vertex x^i for each i in every copy of $K_{1,3}$ in J_n. This yields a dominating set $S = \{x^1, x^2, \ldots, x^n\}$. Also this set is a minimum dominating set in J_n and hence $\gamma(J_n) = n$. This fact is also observed in [3]. Moreover the authors in [3] have determined the $\gamma_t(J_n)$, the total domination number of J_n as $\gamma_t(J_n) = \left\lceil \dfrac{3n}{2} \right\rceil$ if $n \not\equiv 2 \ (mod \ 4)$ and $\gamma_t(J_n) = \dfrac{3n}{2} + 1$ if $n \equiv 2 \ (mod \ 4)$. In this paper, an exact value of the captive domination number of J_n was found. We observe that the total domination set constructed by the authors of [3] cannot be a captive dominating set. That is, the configuration constructed by the authors in [3] and shown in Fig. 3 cannot become a captive dominating set.

According to the authors in [3], when $n \equiv 0 \ (mod \ 4)$, a total dominating set can be constructed by choosing (1) the marked elements in the first four copies of $K_{1,3}$ of J_{4k} as in Fig. 3a; (2) a similar set of marked elements in the 2nd set of four copies of $K_{1,3}$ of J_{4k}, etc.; (k) a similar set of marked elements in the k-th four copies of $K_{1,3}$ of J_{4k}. This means the set S with $6k$ marked elements will constitute a total dominating set. Next, when $n \equiv 1 \ (mod \ 4)$, a total dominating set can be constructed

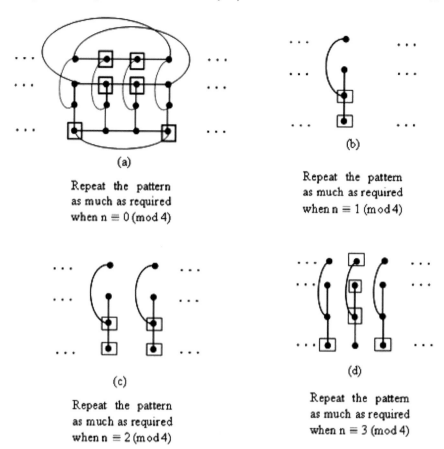

(a)

Repeat the pattern
as much as required
when n ≡ 0 (mod 4)

(b)

Repeat the pattern
as much as required
when n ≡ 1 (mod 4)

(c)

Repeat the pattern
as much as required
when n ≡ 2 (mod 4)

(d)

Repeat the pattern
as much as required
when n ≡ 3 (mod 4)

Fig. 3 The configuration constructed in [3] for $\gamma_t(J_n)$, $n \geq 3$

by choosing the marked elements in the first four copies of $K_{1,3}$ of J_{4k+1} as in the case of J_{4k} and then for the fifth copy of $K_{1,3}$, select the marked vertices as indicated in Fig. 3b. This pattern is repeated for the next successive four copies of $K_{1,3}$ as in Fig. 3a and for the next fifth copy as in Fig. 3b and so on. This will result in the selection of a set S with $6k+2$ marked elements, and this set S will be a total dominating set. Next, when $n \equiv 2$ (mod 4), repeat the procedure said above in Fig. 3a for the first four copies of $K_{1,3}$, and for the subsequent two copies of $K_{1,3}$, the marked elements as in Fig. 3c are selected.

This should be continued for the rest of the next six copies of $K_{1,3}$ and so on. This will result in the selection of a set S with $6k+4$ elements, and this S will be a total dominating set. Finally when $n \equiv 3$ (mod 4) do the same for the first four copies of $K_{1,3}$ as in Fig. 3a and for the subsequent three copies of $K_{1,3}$, select the marked elements as indicated in Fig. 3d and this pattern is repeated for the 7 copies and so on. This will result in the selection of a set S with $6k+5$ marked elements

Fig. 4 The graph J_5

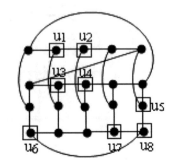

Fig. 5 The graph J_6

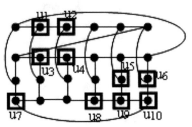

and S will be a total dominating set. In view of this, it follows that $\gamma_t(J_n)$ will be at least $\left\lceil \dfrac{3n}{2} \right\rceil$ for $n \not\equiv 2 \pmod 4$ with $6k, 6k+1, 6k+5$ marked elements when $n \equiv 0 \pmod 4$; $n \equiv 1 \pmod 4$ and $n \equiv 3 \pmod 4$ respectively in order and with $6k+4$ marked elements which is $\dfrac{3n}{2}+1$, if $n \equiv 2 \pmod 4$. However, this process of selection and pattern will not hold good for the computation of captive dominating set when $n \equiv 1, 2 \pmod 4$. This is because, when we search for the marked elements of the pattern chosen in Fig. 3b along with Fig. 3a, we find that some of the marked elements in the constructed total dominating set do not find a neighbor outside S. For instance, consider the first instance of $n \equiv 1 \pmod 4$, namely the graph J_{4k+1} with $k = 1$. Consider Fig. 4. Here $S = \{u_1, u_2, u_3, u_4, u_5, u_6, u_7, u_8\}$.

Even though S is a total dominating set, it is not a captive dominating set as the vertex $u_8 \in S$ has no neighbor in S^c. This anomaly repeats for all $k = 2, 3, \ldots$, in J_{4k+1}. Similarly, consider the first instance of $n \equiv 2 \pmod 4$, namely the graph J_{4k+2} with $k = 2$. Consider Fig. 5.

Here $S = \{u_1, u_2, u_3, u_4, u_5, u_6, u_7, u_8, u_9, u_{10}\}$. Even though S is a total dominating set, it is not a captive dominating set as the vertices u_9, u_{10} are not adjacent with any element in S^c. Hence, we are required to find a different total dominating set that is also a minimal captive dominating set. Mark the elements $w^{2r+1}, x^{2r+1}, x^{2r}, y^{2r}$ for $1 \le r \le k$ in $V(J_{4k+1})$ as the elements of a captive dominating set S. This is because, $(z^s, x^s) \in E(J_{4k+1})$ for $1 \le s \le 4k+1$; $(w^s, w^{s+1}) \in E(J_{4k+1})$ for $1 \le s \le 4k$; $(y^s, y^{s+1}) \in E(J_{4k+1})$ for $1 \le s \le 4k$; $(w^{2r+1}, x^{2r+1}) \in E(J_{4k+1})$ for $1 \le r \le k$; $(x^{2r}, y^{2r}) \in E(J_{4k+1})$ for $1 \le r \le k$ ensuring the dominance, the total dominance and captive dominance properties. So $\gamma_{ca}(J_{4k+1}) \ge 8k+2$. Next, observe that any

Fig. 6 A minimum captive dominating set for J_9

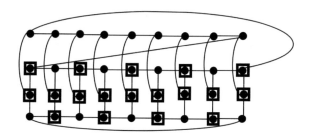

dominating set S of J_{4k+1} must contain x^r for all $1 \leq r \leq 4k + 1$. This is because x^r dominates z^r, w^r and y^r for all $1 \leq r \leq 4k + 1$. As only z^1 and z^{4k+1} are adjacent respectively with w^{4k+1} and w^1 and no other z^r for $2 \leq r \leq 4k$ is adjacent with any w^r for $2 \leq r \leq 4k$, we notice that z^r for $2 \leq r \leq 4k$ cannot become an element of a captive dominating set. Next observe that J_{4k+1} consists of $(4k + 1)$-copies of $K_{1,3}$. As x^r's are included for all $1 \leq r \leq 2k + 1$ in any captive dominating set, we note that each copy of $K_{1,3}$ must contribute at least two elements to the captive dominating set as dictated by the structure of J_{4k+1}. In view of this, if we choose z^1 and z^{4k+1} in the proposed captive dominating set along with x^r for all $1 \leq r \leq 4k + 1$ then in order to satisfy the property of total dominance for the elements x^r for $2 \leq r \leq 4k$, we are compelled to select the elements y^r for $2 \leq r \leq 4k$. But then the middle element y^{2k+1} will not be adjacent with any element outside S and thereby prevents S to become a captive dominating set . So we cannot choose z^1 and z^{4k+1} to be in S and go for the choice of S to be $S = \{x^r$ for all $1 \leq r \leq 4k + 1$; w^r for $1 \leq r \leq k$; y^r for $1 \leq r \leq k\}$ with $|S| = 8k + 2$. As every element in S is indispensable to ensure the captive dominance property, we conclude that $\gamma_{ca}(J_{4k+1}) \leq 8k + 2$ and hence $\gamma_{ca}(J_{4k+1}) = 8k + 2$ (Fig. 6).

On similar lines as above one can claim that $\gamma_{ca}(J_{4k+2}) = 8k + 4$. For the remaining two cases namely when $n \equiv 0 \pmod 4$ or $n \equiv 3 \pmod 4$, the respective dominating sets obtained in [3] will also satisfy the captive dominating set property and hence $\gamma_{ca}(J_{4k}) \geq 6k$ and $\gamma_{ca}(J_{4k+3}) \geq 6k + 5$. Now it remains to obtain reverse inequality in both cases. First let us consider the case J_{4k}. We assert that $6k - 1 = 5$ elements will not suffice for becoming a captive dominating set of J_4. The structure of J_4 demands at least one element to be chosen from each copy of $K_{1,3}$ in J_4.

If each of $K_{1,3}$ in J_4 has one marked element then the fifth element can be from any one of the four copies of $K_{1,3}$. If marked element in the first copy of $K_{1,3}$ is z^1 then the fifth element in the first copy of $K_{1,3}$ must be x^1. In this case then, the marked element in the second copy of $K_{1,3}$ must be z^2 and the marked element in the third copy of $K_{1,3}$ must be z^3 and the marked element in the fourth copy of $K_{1,3}$ must be w^4. This means some of y^is are not even dominated. Again if the marked element in the first copy of $K_{1,3}$ is w^2, then the fifth element in the first copy of $K_{1,3}$ must be x^1 and the marked element in the third copy of $K_{1,3}$ must be w^3 and the marked element in the fourth copy of $K_{1,3}$ must be w^4. In this case also some of the y^is are not even dominated. Again if the marked element in the first copy of $K_{1,3}$ is x^1 then the fifth element in the first copy of $K_{1,3}$ must be y^1. In this case the marked element in the second copy of $K_{1,3}$ must be y^2 and the marked element in the third copy of $K_{1,3}$ must be y^3 and the marked element in the fourth copy of $K_{1,3}$ must be y^4. So some of the z^is are not even dominated. Due to symmetry, the same argument can be repeated if the fifth element falls in the second copy of $K_{1,3}$ or third copy of $K_{1,3}$ or fourth copy of $K_{1,3}$. This shows that $\gamma_{ca}(J_4) \leq 6$ and hence $\gamma_{ca}(J_4) = 6$. Now if $n \equiv 0 \pmod 4$ then the same argument can be repeated for every successive four copies of $K_{1,3}$ of each J_4 and hence $\gamma_{ca}(J_{4k}) = 6k$. Next if $n \equiv 3 \pmod 4$ then consider the case when $n = 7$ or the graph J_7. We claim that $6k + 4 = 10$ elements will not suffice for becoming a captive dominating set of J_7. As in the case of J_4, the structure of J_7 demands at least one element in each copy of $K_{1,3}$ of J_7. If each copy of $K_{1,3}$ in J_7 has one marked element, then the remaining three elements can be distributed in one of the following ways:

(a) All three elements in the first copy of $K_{1,3}$ of J_7 or
(b) Two elements in the first copy of $K_{1,3}$ of J_7 and the third element in the second copy or in the seventh copy of $K_{1,3}$ of J_7 or
(c) One element in the first copy of $K_{1,3}$, one element in the second copy of $K_{1,3}$ in a position adjacent to an already marked element in the first copy and the third element in the third copy or seventh copy of $K_{1,3}$ in a position adjacent to an already marked element in the first copy of $K_{1,3}$. This is all because of the requirement of total domination property.

In the case of (a), all the four elements are marked in the first copy of $K_{1,3}$ of J_7 and in this case some of z^is, w^is, x^is and y^is are not even dominated. In the case of (b), the three marked elements in the first copy of $K_{1,3}$ can be in one of the following four ways. It can be z^1, w^1, x^1 or z^1, w^1, y^1 or z^1, x^1, y^1 or w^1, x^1, y^1. If it is z^1, w^1, x^1, then the tenth element must be either z^2 or w^2 or x^2. But in this case also some of the z^is, w^is, x^is or y^is are not even dominated. If it is z^1, w^1, y^1, then the tenth element must be z^2, w^2, y^2. But in this case also some of the z^is, w^is, x^is, y^is are not even dominated. If it is z^1, x^1, y^1, then the tenth element must be either z^2 or x^2 or y^2. But in this case also some of the z^is, w^is, x^is or y^is are not even dominated. If it is w^1, x^1, y^1, then the tenth element must be either w^2, x^2, y^2. But in this case also some of the w^is, z^is, x^is or y^is are not even dominated. In the case of (c), the already marked element in the first copy of $K_{1,3}$ can be either z^1 or w^1 or x^1 or y^1. The eighth element in that case must be x^1 or y^1. Also the ninth element in the second copy of $K_{1,3}$ must be z^2 or w^2, and the tenth element in the third copy must be z^3 or w^3. In this case also some of the z^is, w^is, x^is and y^is are not even dominated. In a similar manner, we can rule out the other possibilities. Moreover, if the eighth, ninth or tenth element occurs in one of the ways (a) or (b) or (c) in any of the remaining copies of $K_{1,3}$ of J_7, a similar argument can be repeated. This means $\gamma_{ca}(J_7) \leq 11$ and $\gamma_{ca}(J_7) = 11$. Now if $n \equiv 3 \pmod 4$ then the argument can be built up in an identical manner for every successive seven copies of $K_{1,3}$ of each J_7 and hence $\gamma_{ca}(J_{4k+3}) = 6k + 5$. $\square$

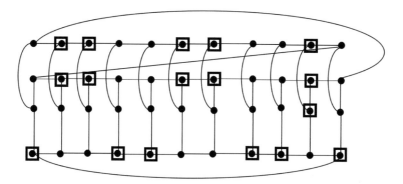

Fig. 7 Flower snark J_{11} graph

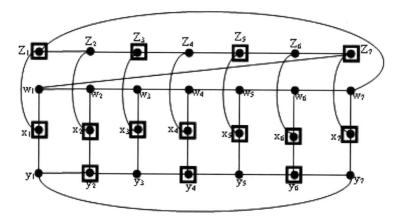

Fig. 8 Flower snark J_7 graph

Note 1 The fact that $\gamma_{sca}(J_n) = \gamma_{ca}(J_n)$ can be seen easily from the definition of strong captive domination number.

Theorem 2 $\gamma_{hcca}(J_{4k+3}) = 8k + 6$.

Proof The fact that $\gamma_{hcca}(J_n) = \gamma_{ca}(J_n)$ can be seen easily from the definition of half certified captive domination number when $n \equiv 1 \pmod{4}$ or $n \equiv 2 \pmod{4}$ or $n \equiv 0 \pmod{4}$. When $n \equiv 3 \pmod{4}$ consider J_{11} shown in Fig. 7.

Observe that every marked element in J_{11} has two neighbors in the marked list of vertices except the marked vertex w^{10}. Hence $\gamma_{hcca}(J_{11}) \neq 17 = 6k + 5$ with $k = 2$ in $4k + 3$ (Fig. 8).

Figure 9 show that $\gamma_{hcca}(J_3) \geq 6$. We claim that 5 elements will not suffice to ensure half certified captive dominance. As marked earlier, each copy of $K_{1,3}$ of J_3 should contain at least one element. Then the fourth and fifth elements can be distributed among the three copies of $K_{1,3}$ in the following ways: (a) Two elements viz., 4th and 5th are in the first copy of $K_{1,3}$ of J_3 (b) 4th element in the first copy

Fig. 9 Flower snark J_3
graph

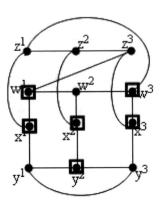

of $K_{1,3}$ and fifth element in the second copy or third copy of $K_{1,3}$ in a position that is adjacent to an already marked element in the first copy of $K_{1,3}$ of J_3. In the case of (a), the three marked elements of the first copy of $K_{1,3}$ can be z^1, w^1, x^1 or z^1, w^1, y^1, z^1, x^1, y^1 or w^1, x^1, y^1. If it is z^1, w^1, x^1, then the already marked element in the second and third copies of $K_{1,3}$ must be z^2 or w^2 and z^3 or w^3. But then in this case some of the y^is are not dominated. If it is z^1, w^1, y^1 then the already marked element in the second and third copies of $K_{1,3}$ must be z^2 or w^2 or y^2 and z^3 or w^3 or y^3. But then in this case some of the z^is, w^is or x^is are not dominated. In a similar manner we can rule out the other possibilities. In the case of (b), the already marked element in the ith copy of $K_{1,3}$ can be z^i or w^i or x^i or y^i for $i = 1, 2, 3, 4$. One can rule out in an exhaustive way all possible combinations by establishing the fact that some of the z^is or w^is x^is or y^is are not even dominated. Hence, $\gamma_{hcca}(J_3) \leq 6$ and $\gamma_{hcca}(J_3) = 6$. This pattern of argument can be extended at every successive J_7, J_{11} ... and hence $\gamma_{hcca}(J_{4k+3}) \leq 8k + 6$. The pattern of marking the elements of J_3, J_7 can be extended to any k and hence $\gamma_{hcca}(J_{4k+3}) = 8k + 6$. $\qquad\square$

Theorem 3

$$\gamma_m(J_n) = \begin{cases} \dfrac{n}{2} & \textit{if } n \equiv 0 \; (mod \; 2) \\ \left\lfloor \dfrac{n}{2} \right\rfloor & \textit{if } n \equiv 1 \; (mod \; 2) \end{cases}.$$

Proof To compute the majority domination number of J_n for any n, we chose the elements x^i for $1 \leq i \leq t$ when $n \equiv 0$ (mod 2) to be the elements of the majority dominating set S and the elements x^i for $1 \leq i \leq \left\lceil \dfrac{2t+1}{2} \right\rceil$ when $n \equiv 1$ (mod 2) to be the elements of the majority dominating set S so that $|N[S]| = 3t + t = 4t \geq \dfrac{4(2t)}{2} = 4t$ when $n \equiv 0$ (mod 2) and $|N[S]| = 3(t + 1) + (t + 1) = 4(t + 1) \geq \dfrac{4(2t + 1)}{2} = 4t + 2$ when $n \equiv 1$(mod 2). This implies that $\gamma_m(J_{2t}) \geq t$ and $\gamma_m(J_{2t+1}) \geq (t + 1)$. If $n \equiv 0$ (mod 2) and $\gamma_m(J_{2t}) < t$ then $|N[S]| = 3(t - 1) + (t - 1) = 4(t - 1) \neq \left| \dfrac{V(J_{2t})}{2} \right| = \left| \dfrac{8t}{2} \right| = 4t$, so a set S with $t - 1$ elements cannot form a majority dominating set . Similarly, if $n \equiv 1 \pmod 2$ and $\gamma_m(J_{2t+1}) < t + 1$ then $|N[S]| = 3t + t = 4t \not\succ \left| \dfrac{4(2t+1)}{2} \right| = \left| \dfrac{V(J_{2t+1})}{2} \right|$ and so a set with t elements cannot form a majority dominating set . So we get a tight bound. That is, $\gamma_m(J_{2t}) \leq t$ and $\gamma_m(J_{2t+1}) \leq t + 1$.

4 Conclusion

We have exactly determined of the captive domination number, strong captive domination number, half certified domination number and majority domination number for Flower Snark graph J_n. We have established a fact that a total dominating set obtained by the authors in [3] cannot directly be carried over to the captive domination number and provided an independent original proof for the captive domination number. It is not known that proof technique provided here could get extended easily to other families of higher order regular graphs. Our effort continues and the outcome of the same will be reported elsewhere. Hence, the problem of determining the various domination numbers discussed here remains open for other families of regular graphs.

Conflict of Interest
The authors declare no conflicts of interest.

Data Availability Statement
This article is self-dependent and does not depend on any data from any other source.

Acknowledgements Yegnanarayanan Venkataraman acknowledges the National Board of Higher Mathematics, Department of Atomic Energy, Government of India, Mumbai, for financial support by their grant no. 02011/10/21NBHM- (R.P)/R&D-II/8007/Date:13-07-2021.

References

1. Abdollahzadeh Ahangar H, Pushpalatha L (2009) Domination number in 4-regular graphs. Int J Math 4:20–30
2. Al-harere M, Omran AA, Breesam AT (2020) Captive domination in graphs. Discret Math Algorithms Appl 12(6)
3. Burdett R, Haythorpe M, Newcombo A (2023) Variants of the domination number for flower snarks. Ars Math Contemp 1:1–28. Received 25th October 2021, Accepted 14th July 2023, Published 31st July 2023
4. Chris Monica M, Sagaya Suganya A (2017) Pebbling in flower snark graph. J Pure Appl Math 13:1835–1843
5. Delbin Prema S, Jayasekaran C (2018) Domination and irredundant number of 4-regular graph. Int J Innovat Res Explorer 5(3):237–241
6. Fraigniaud P, Peters J (2001) Minimum linear gossip graphs and maximal linear (Δ, k)-gossip graphs. Networks 38(3):150–162

7. Kalaiselvi T, Venkatraman Y (2020) On half certified captive domination number of graphs. Accepted for presentation in ICNAAO-2022 & NMD-2022 (IIT Varanasi) and to appear in the Journal of Analysis: Special issue: Nonlinear analysis and optimization
8. Liu H, Sun L (2004) Beijing: on domination number of 4-regular graphs. Czechoslovak Math J 54(129):889–898
9. Maksimovic Z, Kratica J, Savic A, Bogdanovic M (2018) Some static roman domination numbers for flower snarks. In: 13th Balkan conference on operational research, vol 11
10. Mojdeh D, Musawi S, Nazari E, Jafari Rad N (2021) Total domination in cubic Knodel graphs 6(2):221–230
11. Mojdeh DA, Musawi S, Nazari E (2018) Domination in 4-regular Knodel graphs. Open Math 816–825
12. Racicot J, Rosso G (2022) Domination in Knodel graphs 24:1–10
13. da Silva CN, Lucchesi CL (2014) Flower-Snarks are flow-critical. Discret Math
14. Thomasse S, Yeo A (2020) Total domination of graphs and small transversals of hypergraphs. Combinatorica 27(4):473–487
15. Venkatraman Y, Kalaiselvi T, Angelina JJR, Balas VE (2022) A new concept on domination number of graphs, part I. Accept Publ Proc 10th Int Work Soft Comput Appl
16. Xueliang F, Xu X, Yuansheng Y, Feng X (2009) On the domination number of knodel graph $W(3, n)$. IJPAM 50(4):553–558

Design of Microstrip Rectangular Dual Band Antenna for MIMO 5G Applications

M. Jayasudha, P. Ranjitha, Senthil Kumaran R, V. Jayasudhan, and S. Yuvaraj

Abstract In today's wireless communication networks, microwave antennas play a pivotal role in ensuring efficient and reliable connectivity. This is centered on the creation of small rectangular patches for multiband applications using several sorts of flawed soil structure methodologies. This study introduces a proposed design for a microstrip rectangular dual-band antenna specifically tailored for MIMO applications in the context of 5G technology. The antenna is designed to operate at two frequency bands of 3.5 and 6 GHz. The proposed antenna is composed of a rectangular patch with a slit and a rectangular ground plane. The proposed antenna is compact, low-cost, and suitable for 5G MIMO applications. The rectangular patch antenna presented in this study demonstrates a remarkable return loss of -20.02 dB, indicating its excellent impedance matching capabilities. This antenna operates efficiently at a frequency of 5.8 GHz. The suggested antenna performs admirably and has high radiation efficiency. The U-shaped Defected Ground Structure (DGS) employed in this study exhibits an impressive bandwidth of 500 MHz at two central frequencies: 5.7 and 8.8 GHz. Additionally, it achieves a bandwidth of 300 MHz within the frequency range of 8.7–9 GHz. The simulated Frequency Dependent Ground Structure (FDGS) analysis reveals that the 10 dB return loss bandwidth percentage is 5.26%, covering the frequency range of 7.4–7.8 GHz. Moreover, the FDGS achieves an 8.94% bandwidth (11.6–12.7 GHz), demonstrating its effectiveness in providing a wide operating range for the antenna.

The original version of this chapter has been revised: The author name has been corrected. The correction to this chapter can be found at https://doi.org/10.1007/978-981-97-2066-8_19

M. Jayasudha (✉) · P. Ranjitha · Senthil Kumaran R · V. Jayasudhan · S. Yuvaraj
IFET College of Engineering, Villupuram, India
e-mail: jayas1128@gmail.com

Senthil Kumaran R
e-mail: senthilr7@srmist.edu.in

Senthil Kumaran R
SRM Institute of Science and Technology, KTR Campus, Chengalpattu, India

Keywords NOMA · Resource allocation · Sub carrier · Power reduction

1 Introduction

Today is the era of communication; no one can deny this fact. In today's wireless telecommunications market, the traffic is not only voice but also multimedia. Therefore, current communication systems cannot meet the high bandwidth and fast transmission requirements of these multimedia applications. 3G communication systems cannot handle this traffic, so new technologies must be developed for this. The 4G next-generation communication system can be defined as the complete integration of wireless mobile communication and wireless access for multimedia applications. Also, in India, most users today are equipped with "smartphones" and need high-speed internet with all its multimedia applications. However, such applications, such as HDTV on mobile phones, require higher data rates than currently used. To meet the demand for higher data rates, the research team proposed several techniques. The most difficult techniques are directional antenna array and MIMO transmission at the transmitter or receiver. The desired results described above cannot be achieved without proper antenna element design and technology. Many research groups are working on antenna element design, array design, beam direction of arrival estimation, etc. With the advent of smart devices, there is now a need to manufacture smart antennas for these devices. Thick substrates with lower dielectric constants are preferred for achieving optimal antenna performance due to several advantages they offer. These substrates exhibit enhanced efficiency, wider bandwidth, and a looser spatial radiation field. However, it is important to note that these advantages come at the trade-off of bulkier components. For microwave circuits to minimize undesired radiation and coupling, tightly bound fields are necessary. As lower component sizes are desired for microwave circuits, thin substrates with greater dielectric constants are preferred. Ceramic-alumina I ($r = 9.5$, $\tan\delta = 0.0003$): Although minimal, this sort of dielectric loss is delicate. It offers good chemical resistance and high-frequency applications. Alumina can withstand temperatures of up to 1600 °C, but PTFE is a synthetic material ($r = 2.08$, $\tan = 0.0004$). These materials have strong electrical characteristics but weak adhesion and low melting temperatures. This substrate has a rather low level of dimensional stability; however, glass or ceramic reinforcing will make it quite excellent. (iii) Composite-Droid: Composite materials are a combination of the aforementioned synthetic and fiberglass materials ($r = 2.2/6.0/10.8$, $\tan = 0.0017$). These materials have great dimensional stability, and good physical and electrical qualities. (iv) Ferromagnetic ferrite ($r = 9.0$–16.0, $\tan = 0.001$): An electric field polarizes this kind of dielectric. The bias voltage affects the antenna's resonance frequency; (v) semiconductor-silicon ($r = 11.9$, $\tan = 0.0004$): While this dielectric may be incorporated into a circuit, the available area is insufficient for antenna applications. (vi) Fiber Glass—Braided Glass Fiber ($r = 4.882$, $\tan = 0.002$): This substance is quite cheap due to its low loss tangent. Woven fibers, on the other hand, have a tendency to be more anisotropic, which is not ideal in many designs and applications. To encourage radiation-producing fringe fields, the substrate's dielectric constant should ideally be low (r 2.5). Patch antennas are another name for microstrip antennas. Photolithographic techniques are often used to etch the emitter components onto the dielectric substrate. Radiating patches can have any shape, including square, rectangular, circular, oval, triangular, and thin strips (dipoles). Square, rectangular, dipole (strip), and circular geometries have gained popularity in antenna design due to their ease of design and manufacturing as well as their desirable radiation characteristics, notably low cross-polarized radiation. These geometries offer versatility in terms of size and configuration, making them suitable for various applications.

2 Related Works

In 2014, M. Liang, F. Zhang, G. Zhang, and Q. Li, A wideband antenna MICROSTRIP NETWORK with a size of 170×187 mm^2 is designed. To achieve adequate isolation and bandwidth, they positioned two circular planar monopoles (PM) antenna components vertically. The measurements reveal that the microstrip array's antenna impedance bandwidth is 117.6% (0.7–2). Circular antennas for open-coupled feeds were suggested by Deepender Dabas et al. [1] and utilized as feeds for circular channel antennas. For wave communication in the array environment, the aperture coupling overlaps the surface and is closed on all sides of the cavity. To increase heat dissipation, active devices are employed. Active devices are implemented with a pedestal arrangement and a heat sink. 9.3% is the suggested antenna impedance bandwidth (9.55–10.48 GHz). Asem Al-zoubi et al. [2] To achieve an omnidirectional radiation pattern and a wider bandwidth, it is recommended to employ a circular microstrip patch antenna with a substrate-mounted coupling bead ring. This antenna design exhibits a monopole-like radiation pattern, providing coverage in all directions. In comparison to the center-fed circular patch antenna, the suggested design offers a larger bandwidth that better emulates the desired radiation pattern. The suggested antenna design is straightforward to implement, both in terms of its design process and physical construction. It operates within a high impedance bandwidth and delivers a gain of 5.7 dB, ensuring efficient signal transmission and reception. The performance of the circular patch antenna is evaluated through simulations and measurements of return loss and radiation pattern in an anechoic chamber, validating its effectiveness. This type of circular patch antenna finds applications in satellite communication systems and radar systems, where dual-band and wideband capabilities are required. Its versatile nature makes it suitable for a range of wireless communication and sensing applications. Mohammad Sigit Arifianto and others [3]. Proposed a circular patch antenna with a split ring resonator for single-band dual-band identification. A split ring resonator (SRR) has a meta-material structure. Antennas are recommended for 2.4 GHz WLAN and 3.3 GHz WiMAX applications. The circular patch antenna adopts microstrip feed technology and is made of FR4 epoxy resin with a dielectric constant of 4.4 and the height of the substrate is 1.6 mm. The suggested antenna is 55 mm in length and 40 mm in breadth. The suggested circular patch antenna only resonates at 2.4 GHz as opposed to the standard circular patch antenna, proving that it is dual-band capable. The same physical restrictions apply to the manufacture of conventional microstrip antennas on the same dielectric substrate. These two human body antennas are placed side by side to provide the reflection coefficient, VSWR (Voltage Standing Wave Ratio), gain, and radiation pattern. According to the results of the simulation, it is produced by the SRR metamaterial. A circular microstrip patch antenna for dual-band applications featuring a zero-order resonance (ZOR) mode, TM02, is described by Noor Mohammad Awad et al. in their publication [4]. A circular patch antenna is built like a mushroom in a circle. The center-fed circularly polarized antenna has a low profile of 0.02 in the lower frequency bands and a horizontal loop current in the peak plane. The suggested antenna's radiation pattern is discussed. Printed circuit boards with two layers are used to create the suggested antenna. High band: 5. Frequency band; 8–8 dB. 50 ohms through the SMA connection is the recommended input impedance for the antenna feed. The suggested antenna's modeling and measurement results are comparable, and the radiation characteristics of the center-fed circular patch antenna are omnidirectional with two operational frequency bands. Reduce the antenna gain by 5.1 dB and the impedance bandwidth by 0.75%, respectively.

3   Proposed Approach

The proposed antenna consists of a 4 × 4 Rectangular patch Antenna to design the specific operating frequency of the gain, radiation pattern, VSWR, and return loss. For microstrip patch antennas, a formula is available to determine the structure's length and breadth. After knowing the size, dielectric constant, substrate loss tangent, and operating frequency of the chip, the design can be built on software. The designed antenna can then be simulated. Here is the antenna design method used in the proposed works. The rectangular patch antenna's benefit is its small footprint. The Rogers Material-based Patch antenna that has been proposed lacks optical transparency. As a result, the Rogers-based Patch antenna will be used in the upcoming 2–6 GHz band communication architecture. Additionally, a low-power antenna is suggested. The design of a microstrip rectangular dual-band antenna for MIMO 5G applications involves selecting the appropriate dimensions and material for the patch and ground plane, as well as tuning the antenna to the desired operating frequencies. The resonant frequencies of the antenna can be adjusted by changing the length, width, and position of the patch. The following is the design equation for the microstrip patch antenna:

Step 1: Width Calculation (W):The Microstrip patch antenna's width is given as

$$w = \frac{c}{2f\sqrt{\varepsilon r}} = \frac{v_0}{2fr}\sqrt{\frac{2}{\varepsilon\gamma+1}}$$

Step 2: Effective dielectric constant (εreff) calculation:

$$\varepsilon\gamma\,eff = \frac{\in\gamma+1}{2} + \frac{\in\gamma-1}{2}\sqrt{1+\frac{12h}{W}}$$

Step 3: Effective length (Leff) calculation: The effective length is given as

$$Leff = \frac{c}{\sqrt[2f_0]{\varepsilon\gamma\,eff}}$$

Step 4: Length extension (L) calculation: The length extension is given as

$$\Delta L = h(0.412)\frac{(\varepsilon\gamma ef + 0.3)\left(\frac{W}{h}+0.264\right)}{(\varepsilon\gamma ef - 0.258)\left(\frac{W}{h}+0.8\right)}$$

Step 5: Actual length of a patch (L) calculation: The actual length is obtained by

$$2L = \frac{C}{2fr\sqrt{\in\gamma\,eff}} - \Delta L$$

The design microstrip antenna for the 4 × 4 microstrip antenna program uses tone. L&W dimensions are calculated by standard formulas and then an iterative process. Conventional dipole design techniques are used to design a rectangular loop antenna for the proposed frequency of 5.8 GHz. According to the basic antenna theory, the length of a dipole antenna should be half the wavelength, so we first designed a half-wavelength microstrip dipole antenna. Four of these dipole antennas are arranged to form a loop of increasing physical length. The antenna has a straightforward planar construction, and its design and simulation parameters are its effective dielectric constant, which is 4.4 for the FR4 substrate, 1.588 mm for its height, and 0 for its tangent loss. The second iteration of the rectangular dipole 025.is shown in Fig. 1. The length of the structure is 25 mm. The final antenna design size is further reduced to 1/3 of the second iteration, i.e., 25 mm. The width of the dipole is chosen at 2 mm. The real ground planes measurement is 80 mm × 80 mm. The width W and effective length of the Leff structure have been determined to be 72.43 and 80, respectively, 51 mm. The feed for this antenna has been selected as CPW. As seen in Fig. 1, input ports 1 and 2 have @50-Ω impedance SMA connections.

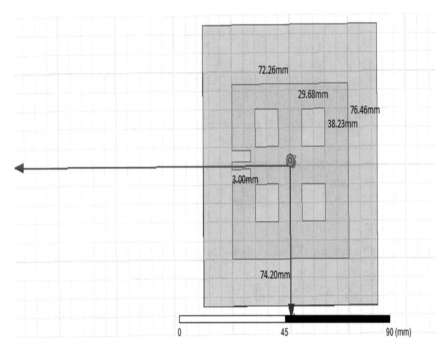

Fig. 1 Antenna design

4 Results and Discussion

The proposed antenna was manufactured and simulated using CST Microwave Studio Version 19. The S-parameters of the antenna were measured and the corresponding results are presented below. The return loss, also known as the reflection coefficient or S11 parameter, was analyzed. The graph below shows the achieved return loss of the proposed antenna, which was −20.00 dB and operates at 5.8 GHz. For the antenna to function properly, the return loss should be less than −10 dB. The effective length of the antenna may be extended while retaining the same antenna area, the design of the microstrip line is crucial. To increase the number of iterations, the effective length is increased and the number of frequency bands is also increased.

To build the microstrip structure, the concept of iteration is used. The Right-Angle Curve was developed and designed by dividing a line into three and replacing the middle section with a curved line segment. These wires are then made into loops to create a new rectangular loop antenna. The lines used to represent the microstrip lines can meander to effectively fill the available space, resulting in an electrically long curve. In terms of bandwidth, effectiveness, and antenna gain, it solves the drawbacks of compact dipole antennas. The technique of moments is the most often used approach for delivering complete wave analysis for microstrip patch antennas. Both differential and integral problems may be solved with it. By multiplying a function by the proper weighting function and integrating, MOM calculates time. Surface currents are used to represent the microstrip patch in a study of a microstrip antenna using this technique, and ground bias currents are used to model the field in the dielectric plate. These integral equations for the electric field are converted to matrix equations using the method of moments, which may then be solved using a variety of algebraic methods to provide the desired answers. Details of the radiation pattern can be understood by knowing the electric dipole charge moment and current distribution in a small segment of the microstrip antenna.

4.1 Return Loss

The simulation results were obtained using a commercially available HFSS. The yield loss achieved is shown in Fig. 2, 7 bands were observed in which the return loss was well below -10 dB. The suggested rectangular patch antenna runsat 5.8 GHz with a return loss of -20 dB. The proposed antenna delivers excellent performance and has a high radiation efficiency.

A commercially available HFSS was used to obtain the simulated results. Figure 2 displays the yield loss that was attained (see Fig. 3).

The proposed rectangular patch antenna has a return loss of -20 dB and operates at 5.8 GHz. The suggested antenna performs admirably and has high radiation efficiency.

4.2 VSWR

VSWR or A measurement of an antenna's impedance matching to the transmission line it is attached to is called the voltage standing wave ratio (see Fig. 4).

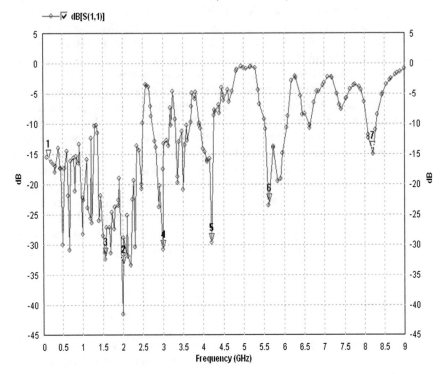

Fig. 2 Return loss obtained by simulation

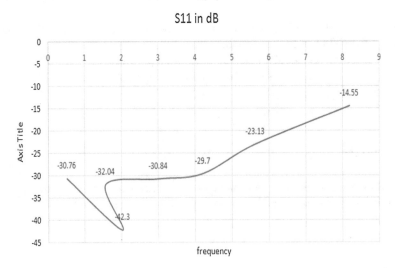

Fig. 3 Graph for resonant frequencies

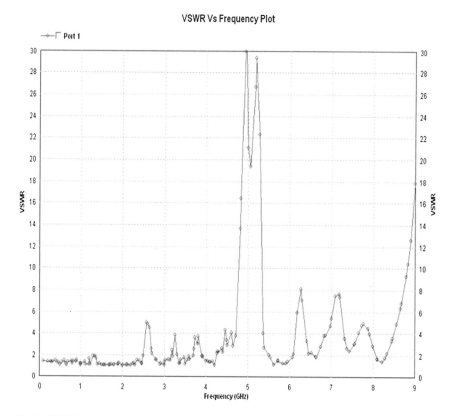

Fig. 4 VSWR obtained by simulation

It is described as the transmission line's highest voltage to minimum voltage ratio, which occurs when there is a standing wave pattern on the line. It achieves a simulated VSWR of fewer than 2 which is 1.9 at 5.8 GHz. The findings are compared to our design and the anticipated multiband behavior of this antenna to support our design. The antenna's axial ratio was found to be zero, demonstrating that it was linearly polarized.

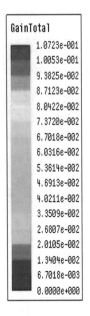

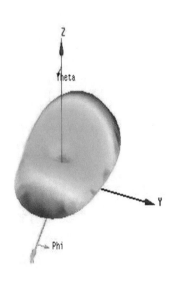

4.3 Gain

The gain of the proposed antenna is about 9.38 dB. The applications of 5.8 Ghz frequency antennas are radio networks, Wireless LAN, ISM applications, etc.

4.4 Radiation Pattern

A radiation pattern in a microstrip patch antenna refers to the directional dependence of the electromagnetic radiation emitted or received by the antenna.

It is a graphical representation of the antenna's response as a function of the angle of incidence of the electromagnetic waves. The proposed microstrip antenna's radiation pattern is shown in Fig. 5.

5 Conclusion

The compact MIMO patch antenna is a versatile and high-performance antenna that finds applications in 5G, mobile communication, vehicular communication, and wireless communication. Using the CST Microwave Studio Suite 2019, we designed a proposed system that operates within the SUB-6 Ghz frequency range. In our

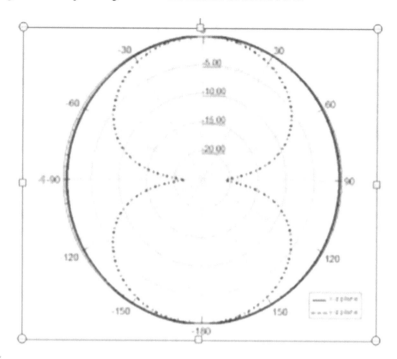

Fig. 5 Radiation pattern

paper, we considered various antenna parameters such as the S11 parameter, VSWR, Directivity, antenna gain, efficiency, and impedance, and compared them with the existing system. Our proposed system outperforms the existing system in terms of VSWR, Return loss, and Directivity. The simulated results show an S11 parameter of −25.09 dB at 4.4 GHz, a VSWR of 1.967 at 4.7 GHz, a Directivity of 6.12 dBi, and an impedance of 50 Ω. The rectangular patches are arranged in a stacking manner. The microstrip patch antenna has a very small bandwidth of just 2–5%. This is an intrinsic restriction of microstrip patch antennas. Patch antennas cannot be utilized for broadband or multiband applications as a result of this restriction. As a result, the multiband patch antenna's design is crucial. Broadband is provided through stacked spotlights. Radar, mobile phones, satellites, UHF RFID, MIMO, and reconfigurable antennas are all examples of modern devices that utilize stacked patch antennas.

References

1. Sahu AK, Misra NK, Mounika K, Sharma PC (2022) Design and performance analysis of MIMO patch antenna using CST Microwave studio, January 2022
2. Vinod Kumar H, Nagaveni TS (2020) Design of microstrip patch antenna to detect breast cancer. ICTACT J Microelectron 6(1):893–896

3. Tung WS, Chiang WY, Liu CK, Chen CA, Rao PZ, Chen SL (2020) Low-cost AIP design in 5G flexible antenna phase array system application. Micromachines 11(9):851–855
4. Mohammed ASB (2019) A review of microstrip patch antenna design at 28 GHz for 5G applications system. Int J Sci Technol Res 8(10):1–14
5. Nayat-Ali O, El Mrabet O, Aznabet M, Khoutar FZ (2019) Phased array antenna for millimeter-wave application. In: Proceedings of IEEE international conference on antennas and propagation, pp 24–25
6. Li J, Zhang X, Wang Z, Chen J, Li Y, Zhang A (2019) Dual-band eight-antenna arraydesign for MIMO applications in 5G mobile terminals, 4 April 2019
7. Paola CD, Zhao K, Zhang S, Pedersen GF (2019) SIW multibeam antenna array at 30 GHz for 5G mobile devices. IEEE Access 7:73157–73164
8. Ahmed Z, Mcevoy P, Ammann MJ (2018) Comparison of grid array and microstrip patch array antennas at 28 GHz. In: Proceedings of IEEE MTT-S international microwave workshop series on 5G hardware and system technologies (IMWS-5G), pp 1–8
9. Yan K, Yang P, Yang F, Zeng L, Huang S (2018) Eight-antenna array in the 5G smartphone for the dual-band MIMO system. In: Proceedings of IEEE international symposium on antennas and propagation and USNC/URSI national radio science meeting, pp 1–12

Classification-Based Credit Risk Analysis: The Case of Lending Club

Aadi Gupta, Priya Gulati, and Siddhartha P. Chakrabarty⬤

Abstract In this paper, we perform a data-driven credit risk analysis, using the data from loan applicants made to a company named Lending Club. The approach adopted in the work required the use of exploratory data analysis and machine learning classification algorithms, namely Logistic Regression and Random Forest. We further made use of the calculated probability of default in order to design a credit derivative (Credit Default Swap) in order to achieve hedging against an event of credit default. The results on the test set are presented using various performance measures.

Keywords Credit risk · Classification algorithm · Exploratory data analysis

1 Introduction

Lending Club, headquartered in San Francisco, was the first peer-to-peer lending institution to offer its securities through the Securities and Exchange Commission (SEC) and enter the secondary market [21]. This article is essentially a case study, on how financial engineering problems can be addressed using Machine Learning (ML) and Exploratory Data Analysis (EDA) approaches. Lending Club specializes in extending different types of loans to urban customers, which is decided on the basis of the applicant's profile. Accordingly, the data considered in this work contains information about past loan applications and whether they were "defaulted" or "not". One of the main objectives of this work is the calculation of the parameters of Probability of Default (PD), Exposure at Default (EAD) and Loss Given Default (LGD), thereby leading to the determination of Expected Loss (EL), making use of

A. Gupta · P. Gulati · S. P. Chakrabarty (✉)
Department of Mathematics, Indian Institute of Technology Guwahati, Guwahati 781039, India
e-mail: pratim@iitg.ac.in

A. Gupta
e-mail: aadi18@alumni.iitg.ac.in

P. Gulati
e-mail: gulati18@alumni.iitg.ac.in

© The Author(s), under exclusive license to Springer Nature Singapore Pte Ltd. 2024
D. Giri et al. (eds.), *Proceedings of the Tenth International Conference on Mathematics and Computing*, Lecture Notes in Networks and Systems 964,
https://doi.org/10.1007/978-981-97-2066-8_8

the dataset under consideration. Also, using the "Recovery Rate" (estimated while calculating "LGD" from "EAD") and the PD, a simple credit derivative is implemented, based on the concept of Credit Default Swaps (CDS) which is used to hedge against such defaults. It may be noted that the total value a lender is exposed to when a loan defaults is the EAD, and the consequent unrecoverable amount for the lender is the LGD [4]. Accordingly, EL is defined as

$$EL = EAD \times LGD \times PD.$$

The driver of this work is the notion of "Classification Algorithm", which weighs the input data (of the applicant, in this case) so as to achieve classifying the input features into two classes, namely positive and negative [9] (default on the loan or not, in this case). Accordingly, we consider two "Classification Algorithms", namely the Logistic Regression and the Random Forest.

The main idea behind Logistic Regression is to determine the likelihood that a particular data point will belong to the positive class (in the case of binary classification) [12]. This is accomplished by associating the independent with the dependent variables via a linear relationship. The weights of the linear relations are determined through the minimization of a Loss Function, which is achieved by using the "Gradient Descent Optimization Algorithm". Since Logistic Regression first predicts the probability of belonging to the positive class, it creates a linear decision boundary (based on a *threshold*, set by the user), separating the two classes from one another. This decision boundary can now be represented as a conditional probability [20]. Implementation of the Random Forest algorithm involves the training stage construction of several decision trees [1], and predictions emanating from these trees are averaged to arrive at a final prediction. Random Forest is considered as an ensemble approach, wherein the Decision Trees are set up so as to achieve optimal partitioning of the datasets into smaller sets for predictive purpose [11, 15]. The evaluation of the model is based on criteria such as Entropy and Gini Impurity. In summary, a decision tree splits the nodes on all the attributes present in the data, and then chooses the split with the most Information Gain, with the Decision Tree model of Classification and Regression Trees (CART) being used in this paper.

The firm-based model uses the value of a firm to represent the event of default, with the default event being represented by the boundary conditions of the process and the dynamics of the firm value [5]. In particular, we refer to two well-established models. Firstly, we mention the Merton model, which is described in [14], which is used to calculate the default probability of a reference entity. In the context, the joint density function for the first hitting times is determined [18]. Secondly, we have the Black-Cox model, which addresses some of the disadvantages of the Merton model [14]. In order to hedge against credit risk, the usage of credit derivatives is a customary approach [3, 7, 13], with CDS being the most common choice of credit derivatives. These types of contracts entail the buyer of the CDS to transfer the credit risk of a reference entity ("Loans" in this case constitute the reference entity) to the seller of the protection, until the credit has been settled. In return, the protection buyer pays premiums (predetermined payments) to the protection seller, which continues until the maturity of the CDS or a default, whichever is earlier [14]. The interested reader may refer to [14] for the formula of CDS spread per annum, to be used later in this paper. Another widely used credit derivative, albeit more sophisticated than the CDS, are the Collateralized Debt Obligation (CDO), which is a structured product, based on tranches [13]. CDOs can further be classified into cash, synthetic and hybrid. The interested reader may refer to [13] for a detailed presentation on pricing of synthetic CDOs. Recent developments in data-driven analytics of credit risk [2] include ensemble-based classifier techniques [17], deep learning-based evaluation model [16] and credit risk analysis using quantum computing [6].

2 Methodology

The goal of this exercise is the approximation of a classification model, on the data considered (from the peer-to-peer lending company, Lending Club), in order to predict as to whether an applicant (whose details are contained in the considered database) is likely to default on the loan or not. Accordingly, to this end, it is necessary to identify and understand the essential variables, and take into account the summary statistics, in conjunction with data visualization. The dataset used for the estimation of PD, EAD, LGD and EL was publicly available on Kaggle [8] and contains the details of all the applicants who had applied for a loan at Lending Club. There were separate files obtained for the accepted and the rejected loans. The file "accepted loans" was only used, since the observations were made on the applicants who ultimately paid the loan and those who defaulted on the loan. The data involved the details of the applicants for the loan, such as FICO score, loan amount, interest rate and purpose of loan. Machine Learning (ML) algorithms were applied on this data to predict the PD after data exploration, data pre-processing and feature cleaning.

The feature selection was performed on the considered data, bearing in mind the goals of predictive modeling. Given the large number of existent features, it was desirable to achieve a reduction in their numbers, thereby retaining only the relevant ones. This action abetted the reduction in computational cost, in addition to improved performance of the model.

The data pre-processing was carried out before the application of the classification model, in order to obtain the probabilities of default. This exercise involved the removal or filling of missing data, removing features which are repetitive or deemed unnecessary and the conversion of categorical features to dummy variables. Some of the dropped columns were "emp_title", "emp_length", "grade", "issue_d" and "title". After pre-processing, the data was partitioned into the customary training and test sets. The former was then used to train the classification model, with the results being evaluated subsequently, on the latter.

In order to accomplish the goal of comparison of different models, the problem of predicting the PD (denoted by p) was treated as a classification problem, where, if $p \geq 0.5$, it is treated as 1 (i.e., the person will default) and if $p < 0.5$, it is treated as

0 (i.e., the person will not default). Accordingly, $p = 0.5$ was used as the *threshold* for the classifier.

An analysis of the data revealed an imbalance between the classes, resulting in the accuracy not being a suitable metric for the comparative analysis across models. Consequently, the approach that we adopted to evaluate the performance was the determination of the "Confusion Matrix", which is a 2×2 combination, namely True Positive/True Negative/False Positive/False Negative (TP/TN/FP/FN). For the purpose of analysis, it is useful to measure two entities, namely *Recall* $\left(\dfrac{TP}{TP + FN} \right)$ and *Precision* $\left(\dfrac{TP}{TP + FP} \right)$, as well as the Harmonic Mean of both (which is called *F1-score*).

Further, the performance evaluation of the classification algorithm can be carried out using Receiver Operating Characteristic (ROC) [19]. ROC being a probabilistic curve, its area under the curve (AUC) reflects the extent of separability and is representative of the extent to which the model achieves classification, with the higher values of AUC indicating better prediction ability of the model [19]. The ROC curve was plotted with True Positive Rate (TPR) (on the y-axis) as a function of False Positive Rate (FPR) (on the x-axis). Note that TPR is synonymous with *Sensitivity* and $1\text{-FPR} = \dfrac{TN}{FP + TN}$ amounts to *Specificity*.

3 Results

The statistics-based feature selection methods were used to assess the correlation between every input variable and the target variable, and those input variables which demonstrated the strongest relationship vis-a-vis the target variable were selected. Further, for selecting the relevant input variables, a comparative analysis was performed, and some of the observations, as a result of this exercise, are noted below

(A) It looks like F and G sub-grades do not get paid back that often.
(B) The loans with the highest interest rates are faced with a greater likelihood of default.
(C) Only 339 borrowers have reported an annual income exceeding 1 million.
(D) It appears that higher "dti" leads to greater likelihood that the loan will be defaulted upon.
(E) More than 40 open credit lines were available to 1211 borrowers.
(F) Finally, the borrower's credit file listed more than 80 credit lines, only in case of 1299 borrowers.

The graph in Fig. 1 shows the correlation between various features, like int_rate and loan_status (i.e., whether the debtor will default or not). Positive correlation implies that more the value of a particular feature, the more is the chance that the debtor will default on their loan.

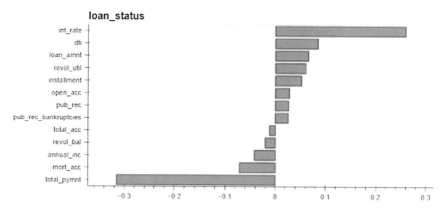

Fig. 1 Correlation Coefficient between Target Variable and Various Attributes

3.1 Logistic Regression

The classification report displays precision, recall and F1-score for both the classes 0 and 1, i.e., applicants who will not default and applicants who will default, respectively. Apart from that, the report also shows Macro Average ($0.5 \times score_0 + 0.5 \times score_1$) and the weighted average ($w_0 \times score_0 + w_1 \times score_1$), where w_0 and w_1 are set contingent on the number of data points belonging to each class. Accordingly, the Accuracy Score obtained is 97.17% and the Classification Report for the "Train Result" is given in Table 1.

Further, the Accuracy Score is 88.81% and the Classification Report for the "Test Result" is given in Table 2.

Table 1 Train evaluation metrics for logistic regression

	0.0	1.0	Accuracy	Macro average	Weighted average
Precision	0.97	0.99	0.97	0.98	0.97
Recall	1.00	0.87	0.97	0.93	0.97
F1-score	0.98	0.92	0.97	0.95	0.97

Table 2 Test evaluation metrics for logistic regression

	0.0	1.0	Accuracy	Macro average	Weighted average
Precision	0.98	0.65	0.89	0.82	0.92
Recall	0.87	0.95	0.89	0.91	0.89
F1-score	0.93	0.77	0.89	0.85	0.90

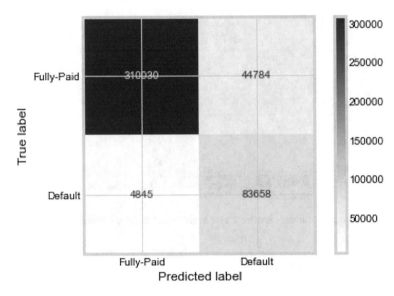

Fig. 2 Confusion matrix for logistic regression

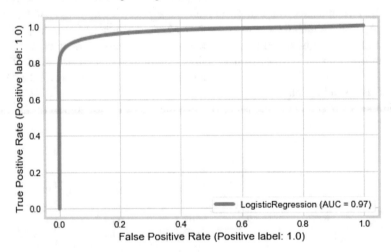

Fig. 3 ROC curve for logistic regression

The Confusion Matrix and the ROC Curve in case of the model using Logistic Regression Model are given in Figs. 2 and 3, respectively.

Table 3 Train evaluation metrics for random forest

	0.0	1.0	Accuracy	Macro average	Weighted average
Precision	1.00	1.00	1.00	1.00	1.00
Recall	1.00	1.00	1.00	1.00	1.00
F1-score	1.00	1.00	1.00	1.00	1.00

Table 4 Test evaluation metrics for random forest

	0.0	1.0	Accuracy	Macro average	Weighted average
Precision	0.97	1.00	0.97	0.98	0.97
Recall	1.00	0.86	0.97	0.93	0.97
F1-score	0.98	0.92	0.97	0.95	0.97

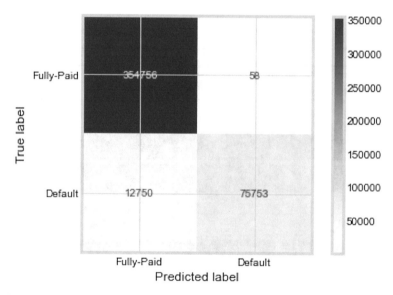

Fig. 4 Confusion matrix for random forest

3.2 Random Forest Classifier

In case of the classifier driven by Random Forest, the Accuracy Score obtained is 100.00% and the Classification Report for the "Train Result" is given in Table 3.

Further, the Accuracy Score is 97.11% and the Classification Report for the "Train Result" is given in Table 4.

The Confusion Matrix and the ROC Curve in case of the model using Random Forest are given in Figs. 4 and 5, respectively.

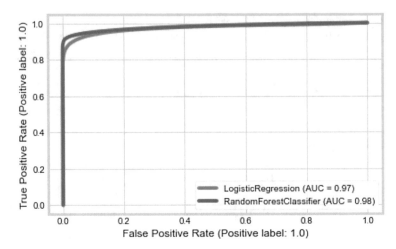

Fig. 5 ROC curve for random forest

Table 5 AUC score for logistic regression and random forest

	ROC AUC score
Random forest	0.928
Logistic regression	0.910

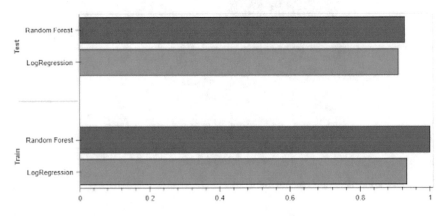

Fig. 6 Graph showing train and test accuracy for both the models

3.3 *Comparison*

Finally, the comparative analysis, based on the AUC, are presented in Table 5 and Fig. 6. It may be noted from Table 5 that, using the AUC score, the performance of Logistic Regression is better than that of Random Forest.

Table 6 Table showing EAD, LGD, EL and derivative price for some (illustrative) customers

Probability of Default	EAD	LGD	EL	Credit derivative price (in bps)
0.09	18330	16727.9	1505.52	0.0469114
0.41	11222.7	10241.8	4199.14	0.156411
0.82	20403.9	18620.6	15268.9	0.447656
0.97	37936.2	34620.6	33581.9	0.38918

3.4 Calculating EAD, LGD, EL and Pricing of Credit Derivative

We begin with the following assumptions in order to calculate EAD and LGD. Recovery rate (which is used to calculate the LGD and the price of the credit derivative) is calculated using the recoveries mentioned for the "Charged Off" clients present in the data [10]. Also, the recovery rate is calculated for each department separately. Pricing of the credit derivative is based upon the theory highlighted in Sect. 1. Further, the following assumptions are made while pricing the derivative:

(A) Only one payment occurs from protection buyer to protection seller (either at the time of maturity or at default time).
(B) The prevalent (risk-free) interest rates are taken into account while calculating the discounting factor.
(C) For the calculation of accrual amount,we assume that on an average, defaults occur right in between a payment period.
(D) At the time of the contract writing, the remaining loan amount (including interest rate) is taken as the notional amount and the loan is treated as a bond.

Table 6 shows some (illustrative) debtors with their probability of default, along with EAD, LGD, EL and the price of the credit derivative.

4 Conclusion

The main objective of the paper was the calculation of probability of default for the debtors of a loan lending organization and the consequent design of a credit derivative (i.e., to calculate the premium) thereby hedging against such defaults. Some conclusions based on the results obtained are as follows:

(A) Based on the performance metrics, it can be concluded that for the data under consideration, the performance of the model based on Random Forest is better than that of the model based on Logistic Regression.
(B) Further, when the dependent and independent variables have a high non-linearity and complex relationship, linear models like Logistic Regression will not per-

form well, since the entire algorithm is based on finding a linear relationship (line of best fit in Logistic Regression) between the variables. In such cases, a model based on Random Forest (non-linear models) will outperform a classical regression method. Hence, there exists a non-linear relationship between the various attributes and the calculated probability.

(C) As expected, the calculated derivative premium is higher for loans linked with obligers with high probability of default, as well as in case of short-term loans.

References

1. Ali J, Khan R, Ahmad N, Maqsood I (2012) Random forests and decision trees. Int J Comput Sci Issues 9(5):272
2. Bhatore S, Mohan L, Reddy YR (2020) Machine learning techniques for credit risk evaluation: a systematic literature review. J Bank Financ Technol 4:111–138
3. Bielecki TR, Rutkowski M (2013) Credit risk: modeling, valuation and hedging. Springer Science & Business Media
4. Bluhm C, Overbeck L, Wagner C (2016) Introduction to credit risk modeling. Chapman and Hall/CRC
5. Duffie D, Singleton KJ (1999) Modeling term structures of defaultable bonds. Rev Financ Stud 12(4):687–720
6. Egger DJ, Gutiérrez RG, Mestre JC, Woerner S (2020) Credit risk analysis using quantum computers. IEEE Trans Comput 70(12):2136–2145
7. Garcia J, Goossens S, Schoutens W (2007) Let's jump together-pricing of credit derivatives: from index swaptions to CPPIS. SSRN 1358704
8. Kaggle: all lending club loan data. https://www.kaggle.com/datasets/wordsforthewise/lending-club
9. Kim H, Cho H, Ryu D (2020) Corporate default predictions using machine learning: literature review. Sustainability 12(16):6325
10. Mora N (2012) What determines creditor recovery rates? https://www.kansascityfed.org/documents/1381/2012-What%20Determines%20Creditor%20Recovery%20Rates%3F.pdf
11. Patel HH, Prajapati P (2018) Study and analysis of decision tree based classification algorithms. Int J Comput Sci Eng 6(10):74–78
12. Peng CYJ, Lee KL, Ingersoll GM (2002) An introduction to logistic regression analysis and reporting. J Educ Res 96(1):3–14
13. Schonbucher PJ (2003) Credit derivatives pricing models: models, pricing and implementation. Wiley
14. Schoutens W, Cariboni J (2010) Levy processes in credit risk. Wiley
15. Sharma H, Kumar S (2016) A survey on decision tree algorithms of classification in data mining. Int J Sci Res 5(4):2094–2097
16. Shen F, Zhao X, Kou G, Alsaadi FE (2021) A new deep learning ensemble credit risk evaluation model with an improved synthetic minority oversampling technique. Appl Soft Comput 98:106852
17. Shen F, Zhao X, Li Z, Li K, Meng Z (2019) A novel ensemble classification model based on neural networks and a classifier optimisation technique for imbalanced credit risk evaluation. Phys A: Stat Mech Its Appl 526:121073
18. Shreve SE (2004) Stochastic calculus for finance II: continuous-time models. Springer
19. Sonego P, Kocsor A, Pongor S (2008) ROC analysis: applications to the classification of biological sequences and 3D structures. Brief Bioinform 9(3):198–209
20. Sperandei S (2014) Understanding logistic regression analysis. Biochem Med 24(1):12–18
21. Wikipedia: lending club. https://en.wikipedia.org/wiki/LendingClub

Lieb Functions and PPT Matrices

Mohammad Alakhrass

Abstract We present several inequalities that govern the components of a 2×2 PPT matrices. The utilization of Lieb functions enables us to present concise and straightforward proofs for these inequalities.

Keywords Lieb functions · PPT matrices · Unitarily invariant norms

1 Introduction

Consider M_n as the set of all $n \times n$ matrices with complex entries. If X and Y as positive definite in M_n, and α belonging to the interval $[0, 1]$, the concept of the geometric mean of X and Y with a weight of α can be described as

$$X \#_\alpha Y = X^{1/2}(X^{-1/2}YX^{-1/2})^\alpha X^{1/2}.$$

If $X, Y \geq 0$, then the weighted mean $X \#_\alpha Y$ is defined as

$$X \#_\alpha Y = \lim_{\epsilon \to 0} X_\epsilon^{1/2}(X_\epsilon^{-1/2}Y_\epsilon X_\epsilon^{-1/2})^\alpha X_\epsilon^{1/2},$$

where $X_\epsilon = X + \epsilon I$ and $Y_\epsilon = Y + \epsilon I$. When $\alpha = \frac{1}{2}$, we just simply write $X \# Y$.
It is known that this binary operation is increasing in the sense that

$$X_1 \#_\alpha Y_1 \leq X_2 \#_\alpha Y_2, \quad \text{when} \quad X_1 \leq X_2 \quad \text{and} \quad Y_1 \leq Y_2. \tag{1}$$

Moreover, we have the inequality:

$$X \#_\alpha Y \leq (1 - \alpha)X + \alpha Y. \tag{2}$$

M. Alakhrass (✉)
University of Sharjah, Sharjah, UAE
e-mail: malakhrass@Sharjah.ac.ae

87

See [3, Chap. 4] for reference.

For $X, Y, Z \in M_n$, let

$$H = \begin{pmatrix} X & Z \\ Z^* & Y \end{pmatrix}.$$

We say that H is positive partial transpose (PPT) if both H and the block $\begin{pmatrix} X & Z^* \\ Z & Y \end{pmatrix}$ are positive semidefinite. The class of PPT has been extensively studied in the literature; for example, refer to [1, 6, 7] and the references therein. Moreover, 2×2 block matrices play a crucial role in the study of matrices, particularly positive matrices [3]. Moreover, they are also very useful in the analysis of sectorial matrices; see for example [2].

In this paper, we provide a proof for the following theorem:

Theorem 1 *Let* $\begin{pmatrix} X & Z \\ Z^* & Y \end{pmatrix}$ *be PPT. Let L be a Liebian function. Then*

$$|L(Z)|^2 \le L\left(X \#_\alpha Y\right) L\left(X \#_{1-\alpha} Y\right), \quad \forall \alpha \in [0, 1].$$

Additionally, we will present several consequences and applications.

Recall that a Lieb function (or Liebian function) L is a continuous function on M_n and satisfies the following characteristics:

1. L is increasing, meaning that for all $X, Y \in M_n$, if $0 < X \le Y$, then $L(X) \le L(Y)$.
2. The following inequality holds true for all $X, Y \in M_n$: $|L(X^*Y)|^2 \le L(|X|^2) L(|Y|^2)$.

For example, the function $L(X) = \text{tr}(X)$, where $\text{tr}(X)$ represents the trace of $X \in M_n$, satisfies the properties mentioned above and hence is a Lieb function. Similarly, the functions determinant, permanent and spectral radius are Lieb functions.

Moreover, unitarily invariant norms serve as prime examples of Lieb functions on M_n. It is important to note that a norm $|| \cdot ||$ is classified as unitarily invariant on M_n if it satisfies the condition $||VXW|| = ||X||$ every $X \in M_n$ and all unitaries $V, W \in M_n$. Here, we provide examples of well-established unitarily invariant norms for $X \in M_n$:

1. Ky Fan k-norms: $\|X\|_{(k)} = \sum_{j=1}^{k} s_j(X)$, where $s_j(X), j = 1, 2, ..., n$, represents the singular values of X, arranged in non-increasing order.
2. Schatten p-norms: $||X||_p = [\text{tr}(|X|^p)]^{1/p}$ for $p \ge 1$. When $p = \infty$, this norm reduces to the usual operator norm defined by $\|X\|_\infty = \sup_{\|X\|=1} \|Ax\|$.

Furthermore, for $X \in M_n$ let $H_k(X) = \prod_{j=1}^{k} \lambda_j(X), X \in M_n$, where $|\lambda_j(X)|$ for $j = 1, 2, ..., n$ represent the eigenvalues of $X \in M_n$ arranged in non-increasing order. For each k, H_k is a Lieb function. Similarly, for each $k = 1, 2, \ldots, n$, the function $F_k(X) = \prod_{j=1}^{k} s_j(X)$, where $s_j(X)$ denotes the singular values of X. These functions are also considered Lieb functions.

2 Proof of Theorem 1

Now, we provide a proof of the main theorem.

Proof It is evident that the condition of $\begin{pmatrix} X & Z \\ Z^* & Y \end{pmatrix}$ being PPT implies that both blocks,

namely $\begin{pmatrix} X & Z \\ Z^* & Y \end{pmatrix}$ and $\begin{pmatrix} X & Z^* \\ Z & Y \end{pmatrix}$, are positive semidefinite. Let us first assume that they are positive definite. Hence, Using Schur's criteria for positive definite matrices, we have $Z^* X^{-1} Z \leq Y$ and $Z^* Y^{-1} Z \leq X$. Next, observe that by the increasing property of means, in (1), we can observe the following:

$$\begin{aligned} Z^*(X \#_\alpha Y)^{-1})Z &= Z^*(X^{-1} \#_\alpha Y^{-1})Z \\ &= (Z^* X^{-1} Z) \#_\alpha (Z^* Y^{-1})Z \\ &\leq Y \#_\alpha X \\ &= X \#_{1-\alpha} Y. \end{aligned}$$

Consequently, we can conclude that $X \#_{1-\alpha} Y \geq Z^*(X \#_\alpha Y)^{-1})Z$. This implies that the block $\begin{pmatrix} X \#_\alpha Y & Z \\ Z^* & X \#_{1-\alpha} Y \end{pmatrix}$ is positive semidefinite for every $\alpha \in [0, 1]$. Now, we utilize the property of Lieb functions, which states that for any Lieb function L and matrices $X, Y, Z \in M_n$, if

$$\begin{pmatrix} X & Z^* \\ Z & Y \end{pmatrix} \geq 0, \tag{3}$$

then we have the inequality $|(Z)|^2 \leq L(X)L(Y)$. See [4, p. 270].
  Applying this result, we obtain

$$|L(Z)|^2 \leq L(X \#_\alpha Y) L(X \#_{1-\alpha} Y), \quad \forall \alpha \in [0, 1],$$

under the assumption that $\begin{pmatrix} X & Z \\ Z^* & Y \end{pmatrix} > 0$ and $\begin{pmatrix} Y & Z \\ Z^* & X \end{pmatrix} > 0$. In the case when $\begin{pmatrix} X & Z \\ Z^* & Y \end{pmatrix}$ and $\begin{pmatrix} Y & Z \\ Z^* & X \end{pmatrix}$ are positive semidefinite definite, we can apply the above argument to the matrices $\begin{pmatrix} X + \epsilon I & Z \\ Z^* & Y + \epsilon I \end{pmatrix}$ and $\begin{pmatrix} Y + \epsilon I & Z \\ Z^* & X + \epsilon I \end{pmatrix}$, where $\epsilon > 0$. This gives us the inequality:

$$|L(Z)|^2 \leq L((X + \epsilon I) \#_\alpha (Y + \epsilon I)) L((X + \epsilon I) \#_{1-\alpha} (Y + \epsilon I)), \quad \forall \alpha \in [0, 1].$$

Now, as we let $\epsilon \to 0$, it implies:

$$|L(Z)|^2 \leq L(X\#_\alpha Y) L(X\#_{1-\alpha} Y), \quad \forall \alpha \in [0, 1].$$

This establishes the desired result.

3 Consequences and Applications

In this section, we present some consequences of Theorem 1.

We start with the following special case.

Corollary 1 *Let* $\begin{pmatrix} X & Z \\ Z^* & Y \end{pmatrix}$ *be PPT. Let L be a Liebian function. Then*

$$|L(Z)| \leq L(X\#Y).$$

Proof By selecting $\alpha = 1/2$ in Theorem 1, we obtain the desired result.

In what follows, we utilize Theorem 1 and consider different types of Lieb functions to establish several inequalities governing the components of PPT blocks. We remark that some of these results are well-known in the literature.

Corollary 2 *Let* $\begin{pmatrix} X & Z \\ Z^* & Y \end{pmatrix}$ *be PPT. Let* $\alpha \in [0, 1]$. *Then*

$$|tr(Z)|^2 \leq tr(X\#_\alpha Y) tr(X\#_{1-\alpha} Y).$$

In particular,
$$|tr(Z)| \leq tr(X\#Y).$$

Corollary 3 *Let* $\begin{pmatrix} X & Z \\ Z^* & Y \end{pmatrix}$ *be PPT. Let* $\alpha \in [0, 1]$. *Then for any unitarily invariant norm we have*

$$||Z||^2 \leq ||X\#_\alpha Y|| \, ||X\#_{1-\alpha} Y||.$$

In particular,
$$||Z|| \leq ||X\#Y||.$$

Corollary 4 *Let* $\begin{pmatrix} X & Z \\ Z^* & Y \end{pmatrix}$ *be PPT. Let* $\alpha \in [0, 1]$. *Then f*

$$\prod_{j=1}^{k} \left|\lambda_j(Z)\right|^2 \leq \prod_{j=1}^{k} \lambda_j(X\#_\alpha Y)\lambda_j(X\#_{1-\alpha} Y).$$

where the eigenvalues are arranged in a decreasing order.

In particular, we have

$$\prod_{j=1}^{k} |\lambda_j(Z)| \le \prod_{j=1}^{k} \lambda_j(X\#Y).$$

As an application of Theorem 1 we derive the following result.

Corollary 5 *Let $X_1, ..., X_m$ and $Y_1, ..., Y_m$ be positive definite in M_n with the property $X_j Y_j = Y_j X_j$ for $j = 1, 2, ..., m$. Then, for all $\alpha \in [0, 1]$ we have*

$$L\left(\sum_{j=1}^{m} X_j^{1/2} Y_j^{1/2}\right)^2 \le L\left(\left(\sum_{j=1}^{m} X_j\right) \#_\alpha \left(\sum_{j=1}^{m} Y_j\right)\right) L\left(\left(\sum_{j=1}^{m} X_j\right) \#_{1-\alpha} \left(\sum_{j=1}^{m} Y_j\right)\right)$$

$$\le L\left(\sum_{j=1}^{m} (1-\alpha)X_j + \alpha Y_j\right) L\left(\sum_{j=1}^{m} \alpha X_j + (1-\alpha)Y_j\right).$$

In particular,

$$L\left(\sum_{j=1}^{m} X_j^{1/2} Y_j^{1/2}\right) \le L\left(\left(\sum_{j=1}^{m} X_j\right) \# \left(\sum_{j=1}^{m} Y_j\right)\right)$$

$$\le \frac{1}{2} L\left(\sum_{j=1}^{m} (X_j + Y_j)\right).$$

Proof The second inequality directly follows from (2), so we only need to prove the first one. Since $X_j Y_j = Y_j X_j$, we have $X_j \# Y_j = Y_j \# X_j = X_j^{1/2} Y_j^{1/2}$. Now, it is clear that the block $\begin{pmatrix} X_j & X_j \# Y_j \\ X_j \# Y_j & Y_j \end{pmatrix}$ is PPT, which implies that

$$\sum_{j=1}^{m} \begin{pmatrix} X_j & X_j^{1/2} Y_j^{1/2} \\ X_j^{1/2} Y_j^{1/2} & Y_j \end{pmatrix} = \begin{pmatrix} \sum_{j=1}^{m} X_j & \sum_{j=1}^{m} X_j^{1/2} Y_j^{1/2} \\ \sum_{j=1}^{m} X_j^{1/2} Y_j^{1/2} & \sum_{j=1}^{m} Y_j \end{pmatrix}$$

is also PPT. Therefore, the result follows from Theorem 1.

We will say that the Lieb function L satisfies the **CM** conditions if it meets the following two conditions:

1. Submultiplicative: For all $X, Y \in M_n$, $L(XY)$ is less than or equal to the product $L(X)L(Y)$.
2. If f is a nonnegative convex function on $[0, \infty)$ such that $f(0) = 0$, then $L\left(\sum_{j=1}^{m} f(A_j)\right) \le L\left(f\left(\sum_{j=1}^{m} A_j\right)\right)$.

An example of such Lieb function is $L(X) = \|X\|$, where $\| \ \|$ is a unitarily invariant norm. This result is presented in [5], it says that if f is a nonnegative convex function on $[0, \infty)$ such that $f(0) = 0$, then for any unitarily invariant norm we have $\| \sum_{j=1}^{m} f(A_j) \| \leq \| f \left(\sum_{j=1}^{m} A_j \right) \|$.

Corollary 6 *Let $X_1, ..., X_m$ and $Y_1, ..., Y_m$ be positive definite matrices in M_n such that $X_j Y_j = Y_j X_j$ for $j = 1, 2, ..., m$. Then, for all $\alpha \in [0, 1]$ and for any Lieb function L satisfies the **CM** conditions we have*

$$L \left(\sum_{j=1}^{m} X_j Y_j \right) \leq L \left(\left(\sum_{j=1}^{m} X_j \right) \#_\alpha \left(\sum_{j=1}^{m} Y_j \right) \right) L \left(\left(\sum_{j=1}^{m} X_j \right) \#_{1-\alpha} \left(\sum_{j=1}^{m} Y_j \right) \right)$$

$$\leq L \left(\sum_{j=1}^{m} (1-\alpha) X_j + \alpha Y_j \right) L \left(\sum_{j=1}^{m} \alpha X_j + (1-\alpha) Y_j \right).$$

In particular,

$$\sqrt{ L \left(\sum_{j=1}^{m} X_j Y_j \right) } \leq L \left(\left(\sum_{j=1}^{m} X_j \right) \# \left(\sum_{j=1}^{m} Y_j \right) \right)$$

$$\leq \frac{1}{2} L \left(\sum_{j=1}^{m} (X_j + Y_j) \right).$$

Proof Since X_j and Y_j are positive commute for each j, we have $X_j Y_j = (X_j^{1/2} Y_j^{1/2})^2$. Therefore, using the fact that $f(x) = x^2$ is convex function and L is a Lieb function and satisfies the **CM** conditions, we can deduce the following:

$$L \left(\sum_{j=1}^{m} X_j Y_j \right) = L \left(\sum_{j=1}^{m} (X_j^{1/2} Y_j^{1/2})^2 \right)$$

$$\leq L \left(\left(\sum_{j=1}^{m} X_j^{1/2} Y_j^{1/2} \right)^2 \right)$$

$$\leq L \left(\sum_{j=1}^{m} X_j^{1/2} Y_j^{1/2} \right)^2.$$

With this, the desired result follows by Corollary 5.

References

1. Alakhrass M (2023) A note on positive partial transpose blocks. AIMS Math V8(10):23747–23755. https://doi.org/10.3934/math.20231208
2. Alakhrass M (2021) On sectorial matrices and their inequalities. Linear Algebr Appl V 617:179–189
3. Bhatia R (2007) Positive definite matrices. Princeton University Press, Princeton
4. Bhatia R (1997) Matrix analysis. Springer, Berlin
5. Kosem T (2006) Inequalities between $f(A + B)$ and $F(A) + f(B)$). Linear Algebra Appl 418:153–160
6. Lee E-Y (2015) The off-diagonal block of a PPT matrix. Linear Algebra Appl 486:449–453
7. Lin M (2015) Inequalities related to 2×2 block PPT matrices. Oper Matrices 9(4):917–924

A Transformer-Based Stock Market Price Prediction by Incorporating BERT Embedding

Parvathi Pradeep, B. Premjith, M. Nimal Madhu, and E. A. Gopalakrishnan

Abstract The stock market trend is known to be volatile, dynamic, and nonlinear. Therefore, accurate prediction of the trend and forecasting the stock prices in today's world is one of the most complex tasks. It is because of the events and preconditions, macro or micro, a few being politics, global economic conditions, and unexpected events which affect the stock market trend. Since it is difficult to predict all the contingencies, how long the effect of such parameters lasts can not be predicted. In this work, we studied the efficacy of different deep learning algorithms to learn the trend in the stock market price to predict the price for the next few days. We considered the stock price, stock index, dollar index, and related news data to predict the stock closing price of Apple Inc. Sentence embedding and sentiment scores were extracted from the news data and fed to the deep learning model along with stock price, stock index, and dollar index values. The deep learning model was designed using a Transformer consisting of an Encoder stack with attention layers and a set of MLP layers to reshape the predictions. The experiments showed that incorporating sentence embedding improved the prediction rate compared to the state-of-the-art model.

Keywords Stock price prediction · BERT · FinBERT · Finance · Transformer

P. Pradeep · B. Premjith (✉) · M. Nimal Madhu
Amrita School of Artificial Intelligence, Amrita Vishwa Vidyapeetham, Coimbatore, India
e-mail: b_premjith@cb.amrita.edu

P. Pradeep
e-mail: cb.en.p2dsc21019@cb.students.amrita.edu

M. Nimal Madhu
e-mail: m_nimalmadhu@cb.amrita.edu

E. A. Gopalakrishnan
Amrita School of Artificial Intelligence, Amrita Vishwa Vidyapeetham, Banglore, India
e-mail: ea_gopalakrishnan@blr.amrita.edu

© The Author(s), under exclusive license to Springer Nature Singapore Pte Ltd. 2024
D. Giri et al. (eds.), *Proceedings of the Tenth International Conference on Mathematics and Computing*, Lecture Notes in Networks and Systems 964,
https://doi.org/10.1007/978-981-97-2066-8_10

1   Introduction

Forecasting stock trends involves thorough knowledge of the market trends for the best returns. It also requires one to be mindful of the risks associated with it. Analyzing the pattern of the stock trend is often based on the sentiment of the overall market, which means that the stock can rise or fall based on the market's conditions or the market dynamics. This dynamic, in turn, depends on various political trends and other economic factors that must be accounted for duly. To capture all these trends, several researchers [4, 6, 7, 16] have included them and have suggested relationships between their changes and the stock trends. One such feature that has a noticeable impact on the stock data is the news articles of such data or the financial data collected from the news. For example, negative news such as a bad earnings report, a lapse in corporate governance, or an unfortunate event typically cause people to sell their stocks and a decrease in the price for many, if not most, stocks. Similarly, positive news such as good earnings, the announcement of a new product, and acquisitions can indicate an increase in stock price and more purchases of stocks.

The sentiments derived from the news can have an observable effect on the investor's choices [12]. They play a significant role in forecasting stock prices. Specific keywords can bring out the overall sentiment in the news heading that reflects the market's current situation. This aspect is utilized in several researches [1, 3, 8, 11, 18] to determine the stock trends, and they have concluded that sentiments strongly impact the trend flow.

Adding news to the stock data can help diminish the highly random volatile nature of the prediction of the stock prices as it can help to anticipate and adjust the next move we choose in investing. There is a direct correlation between the trading prices from a group of stocks and the sentiment the news portrayed. For processing the data extracted from news, Google's Bidirectional Encoder Representations from Transformers model (BERT) [5, 9] is a state-of-art model for obtaining textual embedding and sentiment classification.

The Financial version of the BERT or FinBERT model is considered in this work as it is pre-trained with data-rich financial jargon and can help it capture the financial sentiments in the new tag lines better than standard BERT. The sentence embedding and the sentiment scores of the text ranging from -1 for negative sentiment to 0 for neutral and +1 for positive sentiment are extracted to be added into the dataset along with the classic OHLC (Open, High, Low, and Closed price) of the data.

Dimension reduction of the generated embedding is necessary to accommodate the stock indices. The dimension reduction should ensure that most of the information from the original data is retained—the storage issue for higher dimension embedding increases when there is a large volume of data involved. With dimension reduction methods, the computation costs are reduced. New features will be created by combining information from the original features that capture much of the information contained in the original features. There are also linear and nonlinear transformations, depending on the combining technique applied. Some methods for dimension reduction include Principal Component Analysis, AutoEncoders [19, 21],

UMaps, etc. PCA is used for linear transformations, and deep neural networks like AutoEncoders and UMaps are used for linear and nonlinear transformations. In this thesis, the reduction using PCA and AutoEncoders was compared.

The use of embedding here can help get a better sense of the semantic information in the given text, and the addition is expected to improve the recognition of the trend or pattern of the stock price. The embedding is generated with the help of several models, such as the Finbert-tone model [10], DistilBERT model [17], DistilRoBERTa base [17] model, and the Bert base uncased models [5].

Training on the traditional time series models can help to get a general idea of what the model understands from the data. In this study, we used a Transformer network consisting of an Encoder stack consisting of several Multi-head attention layers and a feed-forward network. We added MLP layers after the stack was used to reshape the predictions for the timestep provided. The motivation for using Transformers in this work is to predict the stock prices given the actual embedding of the news as an input. This model was compared to variations of GAN models that could generate stock prices similar in distribution to the actual data.

This work prioritizes the inclusion of semantic structure of the news heading for forecasting the stock price. The embedding taken from the texts places semantically similar inputs close together in the embedding space, which can help identify a pattern. This information, along with the sentiment score calculated from the news, was used in determining the trend of the stocks for an upcoming set of days ahead. Adding embedding is advantageous as the models can learn and reuse it. The embedding was generated for the whole sentence using sentence transformer [15], and the sentiments were calculated with the help of the FinBERT model [10]. The models used in this study include LSTM, RNN, and GRU [14], along with the AutoEncoders and the Transformer model.

2   Literature Review

Stock market predictions are generally made with the help of time series forecasting models such as LSTM, GRU, and ARIMA. [2] However, some studies make stock predictions using GAN, as GAN is generally used to process 2D data or images. In paper [22], the authors used a comprehensive method of using LSTM, GRU models as generators and CNN models as discriminators in the GAN architecture and have improved its performance by regularizing the loss using the Wasserstein distance as loss function and using Gradient Penalty method to tune the model. The data used included the before and after pandemic data, which can help determine if the model could capture the contingencies in the trends. The results showed that while the GAN model worked well with pre-pandemic data, the WGAN-GP model worked best for the data, including the pandemic time. In the case of transformers, [20] replaces the LSTM and CNN models with an attention mechanism and works successfully for NLP problems such as translations, sequence predictions, and questions answering. In the paper [13], the authors have experimented with Transformers on stock data and have used the sequential models as baseline models for comparison. This paper has used a dataset containing 11 features, including technical stock indicators like moving averages. The results were better than the baseline models, with a directional accuracy of 0.71 and 0.89 for larger time lags. The predicted days' window is smaller, i.e., 2 weeks or 3 weeks.

3   Objectives

The main objective of this work is to identify the algorithms that can best identify the stock trends given the data and then forecast the stock price for the next few days by including the features extracted from related news headlines. In order to achieve this, financial news must be extracted from reliable sources. The dataset input to the task-specific transformer will have the dimension-reduced embedding and predict prices for different timesteps. The final step of this work includes comparing the performance of all the models by training and testing with the data from both before and after the COVID-19 pandemic.

4   Methodology

This section discusses the methodology proposed in this paper. Here, the dataset has two parts. The first part is formed by combining stock indicators, dollar indices, and the OHLC data. The second part contains the financial news related to the stock collected daily. It was achieved by scrapping the data from the daily news of Apple Inc. from Seeking Alpha.[1] The next step involves extracting relevant features from the text (or news) data. Here, we considered two features to be extracted from the news headlines—embedding the headline and the degree of sentiment expressed in the headline. We considered different BERT-based models for generating the embedding, whereas the sentiment score was extracted using the FinBERT model, which is trained exclusively for financial text data. The embedding dimension generated by the BERT models was 1×768. In order to reduce the dimensions, both the dimension reduction methods: PCA and Auto Encoders were experimented with, and the latter gave improved results. The Auto Encoders are trained to learn so that the output equals input. It generally consists of an encoder and decoder stack. The encoder ($f(.)$) maps the input to an intermediate embedding representation (h) called the bottleneck, and the decoder ($g(.)$) maps back to the input space using the representation h.

---

[1] https://seekingalpha.com/.

The equations for encoder and decoder, respectively, will be as follows.

$$h = f(W_h x_i + b_h)$$
$$\hat{x}_i = g(W_o h + b_o) \tag{1}$$

where i and o is for input and output and W and b is for the weights and bias, respectively.

The dimension of the embedding reduced with the help of Auto Encoders was 64 to match the size with the other features. It helps to avoid the model being biased toward sentence embedding instead of considering all the features equally. The final dataset consisted of the OHLC features, the news tagline vector representation, and the FinBERT score combined to form a 70 feature vector.

The data has to be modeled by sequential learning algorithms to forecast the stock closing price. Sequential algorithms such as RNN, LSTM, and GRU were used in this work.

In the next model, we used transformer architecture for the prediction. In the original Transformer paper by [20], the encoder stack consists of 6 identical layers. Each layer is composed of a multi-head attention mechanism followed by a fully connected feed-forward network. The multi-head self-attention is preferred over the self-attention mechanism as it calculates the importance of each token to the other tokens in the sequence, therefore, creating a more dynamic representation. In each of the attention layers in the Multi-head Attention layer, three mappings, i.e., the query, key, and value, from the embedding are learned. Using each of the token keys and all token query vectors, a similarity score is calculated with the dot product. These scores are used to weigh the value vectors to arrive at a new representation of the token. With Multi-head Attention, these scores will be concatenated together for representing different perspectives with which the word can be identified from the embedding vector. The following are the equations for the Multi-Head attention layers.

$$Q^{(h)}(x_i) = W_{h,q}^T x_i$$
$$K^{(h)}(x_i) = W_{h,k}^T x_i$$
$$V^{(h)}(x_i) = W_{h,v}^T x_i \qquad W_{h,q}, W_{h,K}, W_{h,V} \in \mathbb{R}^{k \times \daleth} \tag{2}$$
$$\alpha_{i,j}^{(h)} = softmax_j \left(\frac{\langle Q^{(h)}(x_i), K^{(h)}(x_j) \rangle}{\sqrt{k}} \right)$$

Here, $Q^{(h)}(x_i), K^{(h)}(x_i)$, and $W^{(h)}(x_i)$ represent the query, key, and value vector for each of the head h in Multi-Head Attention layer. The weights $\alpha_{i,j}^{(h)}$ are referred to as attention weights, which control how much element x_i "attends" x_j in head h.

The Transformer model used in this work will consist of an Encoder stack, which has m number of Multi-Head Attention layers along with a feed-forward network. The hyperparameters include the number of Multi-head layers, m, the number of encoder blocks in the stack, the hidden layer size in the feed-forward network, and so on. A polling layer is added to aggregate the outputs of the transformer stack,

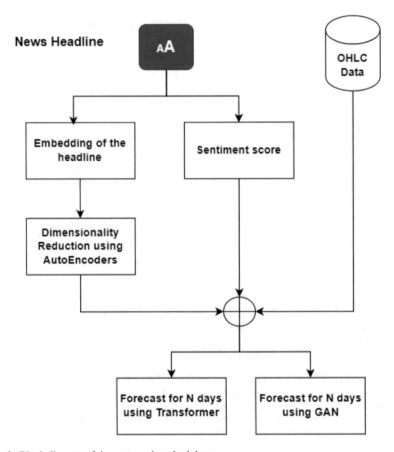

Fig. 1 Block diagram of the proposed methodology

followed by an MLP unit with n dense layers to reshape the outputs as per the given timestep. In the case of transformers, they had performed well with lower timesteps.

The metric to identify the performance will be the RMSE score based on the predicted sequence generated by the transformer model and the actual data. The final steps include the comparison of results between the model trained with the pre-pandemic data alone and the data with both pre-pandemic and pandemic data (Fig. 1).

5 Dataset Description

The dataset comprises stock indices that are extracted from Yahoo Finance, dollar indices from Fred, and the news data is scrapped from the stock market news website,

Seeking Alpha. The target variable is the closing price of Apple's stock, i.e., the forecasting will be done on it. The data is collected for a period of 10 years (2010–2020) and consists of around 6000 observations with 36 primary features, which are further reduced to the OHLC features. The sentiment score obtained from the news is also added to the dataset. Finally, the embedding of the news headlines, generated by the BERT model, is dimensionally reduced to 64 columns. Thus, there are 70 columns in total, including the target variable. The dataset will be split for train, validation, and test depending on the time period, i.e., train and validation samples are from the pre-pandemic period (2010–2019) and there are two sets of test datasets; one from the pandemic period (2020) and the other post-pandemic.

6 Experimentation Results and Discussions

The dataset with the stock indices and the news embedding are split according to the date, i.e., from 2010 to 2019 is considered for training (5983 data points), and the rest (255) data points are considered for training with pandemic data. The total number of features is 70 which includes the closing price as the target variable.

In this work, the RMSE is taken as the metric for determining the model's performance which is calculated as follows:

$$RMSE = \sqrt{\frac{\sum_{i=1}^{N}(x_i - \hat{x}_i)^2}{N}} \qquad (3)$$

where N is the number of data points, x_i is the actual or real data, and $\hat{x}_i$ is the predicted data.

In the initial experiments, we utilized five deep neural networks to determine the generator for the GAN. These networks included RNN, LSTM, GRU, Bidirectional LSTM, and Stacked LSTM. Except for the stacked LSTM, all networks had a single hidden layer with 128 neurons. The stacked LSTM had three hidden layers with 128, 64, and 32 neurons in the respective layers. The batch sizes were set to 8, 16, and 32 during training, which lasted for 20 epochs and resulted in converged loss for all experiments. The models' performance was evaluated using the RMSE metric. Among the models, the RNN model with DistillBERT-base embedding demonstrated the best performance. While the GRU model showed promise during training and validation, it underperformed during testing. Both Bi-LSTM and Stacked LSTM models exhibited poor performance, with considerable loss.

Table 1 Auto Encoder versus PCA with GRU and RNN

Sr. no	Model	Timestep	RMSE score with PCA	RMSE score with Auto Encoder
1	GRU	10	3.3	2.7
2		30	3.6	3.4
3		50	$3.1E + 04$	2.8
4	RNN	10	3.36	1.7
5		30	4.2	2.3
6		50	2.8	1.4

6.1 Dimension Reduction

In order to reduce the DistillBERT-base embedding dimension, PCA and Auto Encoders were used. The embedding of size $n \times 768$ is reduced to a dimension of $n \times 64$, n being the number of rows in the dataset. Then they were tested with the sequential models to compare their performances. The results showed that Auto Encoders performed better than PCA. This is because the Auto Encoders could capture both the linear and nonlinear relations in the data.

In the AutoEncoders used for dimension reduction, the number of neurons decreased as the layers increased in the case of encoders and the neurons increased as the layers increased in the case of decoders, i.e., they were symmetrical in shape. The reduced dataset will be extracted from the bottleneck region.

The number of neurons in the encoder and decoder was a hyperparameter that was decided based on the experiments conducted. The dimension reduction of the embedding was done with both PCA and Auto Encoders and its results were checked on the sequential models (GRU,RNN). The input had timestep 10, 30, and 50 for reference (Table 1).

The results from the Auto Encoder were significantly better than that of PCA. This shows that the Auto Encoder can easily outperform the PCA when the data is nonlinear.

6.2 Transformers for Stock Prediction

The Encoder model considered in this paper consists of a stack having the m layers of Multi-Head Attention layers followed by the feed-forward network with the layer normalization and the residual skip. The structure of the model is shown in Fig. 2.

The hyperparameters used in the transformer model include the timestep, embedding size for attention model, hidden layers in the feed-forward network, and the MLP units used. The transformer model performs well with lower timestep in order to achieve improved parallelizability and better parameter sharing of weights. So

Fig. 2 Block diagram of
transformer model

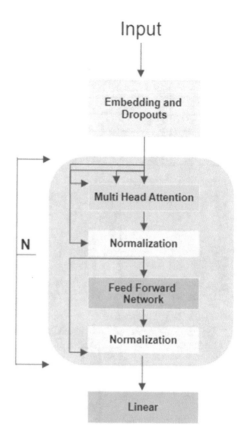

timesteps lower than 10 were also considered in the experiments. The timestep above 10 day considered for transformers, i.e., 30 and 50 gave poor results. The timestep was further reduced to check the performance of the models. The experiments were conducted on two types of data. One was with a separate train and test set which is split as per the year, i.e., the training set with data from the year 2010–2019 and the test set with the data from the year 2020 (the year the COVID-19 pandemic hit). The second type is with the train and test set combined and then split. This was to check for any bias in results based on the year (Table 2).

The second set of experiment has the train and test set combined in an 80∶20 ratio. This ratio was decided after running multiple ratios with the same timestep. The following set of graphs in Figs. 3 and 4 show the results of the transformers for different timesteps and for both the combinations of dataset.

The above results were compared with that from the base paper [10] as shown in Fig. 3, and the results were comparatively better. The reference paper had an RMSE score of 4.77 in case of WGAN model with data including the year 2020 and 3.88 testing with data excluding the year 2020 (Table 3).

Table 2 Evaluation of Transformer with RMSE score for Train and Test separately and combined

Timestep	RMSE score with train and test set separate	RMSE score with train and test set combined
2	1.86	2.76
3	6.3	3.89
5	4.71	5.06
10	2.75	5.66
30	6.78	9.78
50	5.68	10.63

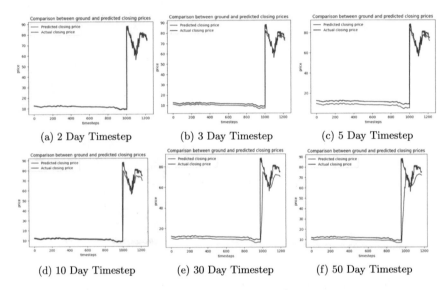

(a) 2 Day Timestep   (b) 3 Day Timestep   (c) 5 Day Timestep

(d) 10 Day Timestep   (e) 30 Day Timestep   (f) 50 Day Timestep

Fig. 3 Results with train and test set given separately in transformer

Table 3 RMSE score comparison with base paper

Sr. no	Paper	Model	Sample set	RMSE score
1	Benchmark models	GAN	Training set	3.09
2		WGAN	Training and test set	4.77
3	Proposed word model	Transformer	Training set	1.86
4		Transformer	Training and test set	2.76

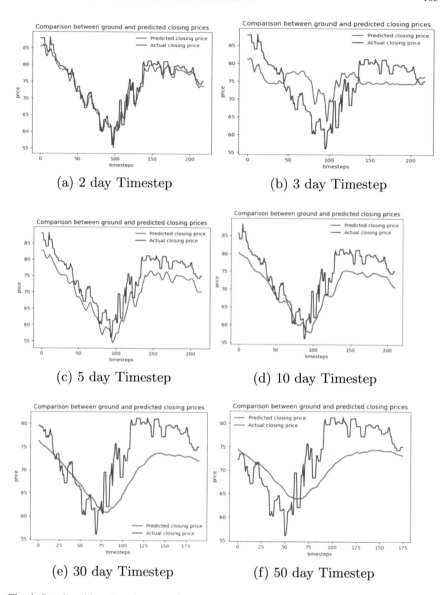

Fig. 4 Results with train and test set given combined in transformer

7 Conclusion

This work discusses using Transformers for forecasting the stock prices for different timesteps by incorporating the features extracted from corresponding news headlines. Sentence embedding of the headlines and sentiment scores of the headlines were considered as the text features. Distillbert-case and FinBERT models were used for generating the sentence embedding, and FinBERT was used for computing the sentiment scores. The embedding was reduced in shape using Auto Encoders and the forecasting was done with the help of Transformers. Different hyperparameters and variations of the data were tested on the transformers. The inference that can be made from these experiments is that as the number of days increases, the loss for prediction (RMSE loss) also increases. It had the opposite relationship when tested with the variation of GAN models, which were constructed based on the reference paper [10]. The inference that can be made is that the transformer model gave better results with the lowest timestep, i.e., 2. The dataset with the train and test set split based on year gave the lowest RMSE score of 1.86. In addition, the embedding and the sentiment scores obtained from the news headlines improve the forecasting rate substantially.

The results can be varied when considering more data. Furthermore, the prediction sequence can be taken minutely and experimented with, provided consistent data is available. An addition of deep reinforcement learning techniques can also be used to improve further and stabilize the network's performance for prediction.

References

1. Ashtiani MN, Raahemi B (2023) News-based intelligent prediction of financial markets using text mining and machine learning: a systematic literature review. Expert Syst Appl 217:119509. https://doi.org/10.1016/j.eswa.2023.119509. https://www.sciencedirect.com/science/article/pii/S0957417423000106
2. Balaji AJ, Ram DH, Nair BB (2018) Applicability of deep learning models for stock price forecasting an empirical study on bankex data. Proc Comput Sci 143:947–953
3. Bi J et al (2022) Stock market prediction based on financial news text mining and investor sentiment recognition. Math Probl Engin 2022
4. Dahal KR, Pokhrel NR, Gaire S, Mahatara S, Joshi RP, Gupta A, Banjade HR, Joshi J (2023) A comparative study on effect of news sentiment on stock price prediction with deep learning architecture. Plos one 18(4):e0284695
5. Devlin J, Chang MW, Lee K, Toutanova K (2019) BERT: pre-training of deep bidirectional transformers for language understanding. In: Proceedings of the 2019 conference of the North American chapter of the association for computational linguistics: human language technologies, vol 1 (Long and short papers). Association for Computational Linguistics, Minneapolis, Minnesota, pp 4171–4186. https://doi.org/10.18653/v1/N19-1423. https://aclanthology.org/N19-1423
6. Gite S, Khatavkar H, Kotecha K, Srivastava S, Maheshwari P, Pandey N (2021) Explainable stock prices prediction from financial news articles using sentiment analysis. Peer J Comput Sci 7:e340

7. Kalyani J, Bharathi P, Jyothi P et al (2016) Stock trend prediction using news sentiment analysis. arXiv preprint arXiv:1607.01958
8. Kaya M (2010) Karsligil M. Stock price prediction using financial news articles 10:478–482. https://doi.org/10.1109/ICIFE.2010.5609404
9. Kumar CA, Maharana A, Murali S, Premjith B, Kp S (2022) Bert-based sequence labelling approach for dependency parsing in tamil. In: Proceedings of the second workshop on speech and language technologies for dravidian languages, pp 1–8
10. Lin H, Chen C, Huang G, Jafari A (2021) Stock price prediction using generative adversarial networks. J Comput Sci 17–188
11. Mohan S, Mullapudi S, Sammeta S, Vijayvergia P, Anastasiu DC (2019) Stock price prediction using news sentiment analysis. In: 2019 IEEE fifth international conference on big data computing service and applications (BigDataService), pp 205–208. https://doi.org/10.1109/BigDataService.2019.00035
12. Nair BB, Kumar P, Prasad S, Singh L, Vijayalakshmi K, Sai Ganesh R, Reshma J (2016) Forecasting short-term stock prices using sentiment analysis and artificial neural networks. J Chem Pharmaceut Sci 9(1):533–536
13. Nguyen TH, Shirai K, Velcin J (2015) Sentiment analysis on social media for stock movement prediction. Expert Syst Appl 42(24):9603–9611
14. Premjith B, Soman K (2021) Deep learning approach for the morphological synthesis in malayalam and tamil at the character level. Trans Asian Low-Resour Lang Inf Process 20(6):1–17
15. Reimers N, Gurevych I (2019) Sentence-bert: Sentence embeddings using siamese bert-networks. In: Proceedings of the 2019 conference on empirical methods in natural language processing. Association for computational linguistics. https://arxiv.org/abs/1908.10084
16. Ren Y, Liao F, Gong Y (2020) Impact of news on the trend of stock price change: an analysis based on the deep bidirectiona lstm model. Proc Comput Sci 174:128–140
17. Sanh V, Debut L, Chaumond J, Wolf T (2019) Distilbert, a distilled version of bert: smaller, faster, cheaper and lighter. arXiv:abs/1910.01108
18. Shah D, Isah H, Zulkernine F (2018) Predicting the effects of news sentiments on the stock market. In: 2018 IEEE international conference on big data (Big Data). IEEE, pp 4705–4708
19. Todo W, Laurent B, Loubes JM, Selmani M (2022) Dimension reduction for time series with variational autoencoders. arXiv preprint arXiv:2204.11060
20. Vaswani A, Shazeer N, Parmar N, Uszkoreit J, Jones L, Gomez AN, Kaiser Ł, Polosukhin I (2017) Attention is all you need. In: Advances in neural information processing systems, pp 5998–6008
21. Wang Y, Yao H, Zhao S (2015) Auto-encoder based dimensionality reduction. Neurocomputing 184:11. https://doi.org/10.1016/j.neucom.2015.08.104
22. Zhang K, Zhong G, Dong J, Wang S, Wang Y (2019) Stock market prediction based on generative adversarial network. Proc Comput Sci 147:400–406

On the Solution Set of Semi-infinite Tensor Complementarity Problem

R. Deb and A. K. Das

Abstract To give a method for thinking about a more practical scenario of the problem, we introduce the semi-infinite tensor complementarity problem in this study. We provide the sufficient and necessary conditions for the solution set's existence. We examine the solution set's error bounds in this case using the residual function.

Keywords Semi-infinite tensor complementarity problem · Residual function · Error bound · Feasible solution · R_0-tensor

1 Introduction

Song and Qi introduced the tensor complementarity problem, $\text{TCP}(q, \mathcal{A})$, in their publications [32, 33]. In recent years several theoretical as well as computational research works have been done on this topic. Multilinear game problem is one of the crucial issues in optimization theory. By providing a reformulation of the multilinear game as a tensor complementarity problem, Huang and Qi [12] established a connection between the multilinear game problem and the tensor complementarity problem. They illustrated how resolving the ensuing TCP is equal to identifying the multilinear game's Nash equilibrium point.

Given two positive integers m and n, we consider $\mathcal{A}$ as an n dimensional real tensor of order m, i.e., $\mathcal{A} \in T_{m,n}$. When a real vector $q \in \mathbb{R}^n$ is supplied, the objective of the tensor complementarity problem, represented as $\text{TCP}(q, \mathcal{A})$, is to determine a real vector $x \in \mathbb{R}^n$ for which the following conditions hold.

$$q + \mathcal{A}x^{m-1} \geq 0, \quad x \geq 0, \quad \text{and} \quad x^T(q + \mathcal{A}x^{m-1}) = 0. \tag{1}$$

R. Deb (✉)
Jadavpur University, 188 Raja S.C. Mallik Road, Kolkata 700 032, India
e-mail: rony.knc.ju@gmail.com

A. K. Das
Indian Statistical Institute, 203 B. T. Road, Kolkata 700 108, India
e-mail: akdas@isical.ac.in

© The Author(s), under exclusive license to Springer Nature Singapore Pte Ltd. 2024
D. Giri et al. (eds.), *Proceedings of the Tenth International Conference on Mathematics and Computing*, Lecture Notes in Networks and Systems 964,
https://doi.org/10.1007/978-981-97-2066-8_11

109

Consider $m = 2$, i.e., the order of the tensor is two. Then the problem (1) gets reduced to the well-known linear complementarity problem. Given a real matrix A of order $n \times n$ and a real vector q in $\mathbb{R}^n$, the linear complementarity problem is indicated as LCP(q, A). The objective of the linear complementarity problem is to determine a real vector $x \in \mathbb{R}^n$ for which the following conditions hold.

$$q + Ax \geq 0, \quad x \geq 0, \quad \text{and} \quad x^T(q + Ax) = 0. \tag{2}$$

A wide range of optimization issues are taken into account by the complementarity concept. Linear programming problem, linear fractional programming, convex quadratic programming problem, and the bimatrix game problem are among the issues that can be reformulated as linear complementarity problems. Problems with linear complementarity framework are extensively discussed in the literature on mathematical programming and have several applications in engineering, geometry, mathematical economics, operations research, and control theory. For current works on this subject and its applications, see [6, 9, 14, 16, 18–20, 23, 24, 26] and the references cited therein.

The widely recognized linear complementarity problem served as the initial inspiration for the PPT notion, which is now used in numerous different contexts. In essence, the PPT is a matrix that transforms a linear system used to exchange unknowns with matching entries on the system's right side. For further information, see [3, 5, 21, 25, 30].

In the complementarity theory with a linear framework, several matrix classes and their subclasses have received substantial investigations. These studies of matrix classes are done due to their prominence in complexity theory, scientific computing, and the theoretical underpinnings of linear complementarity problems. For current works on this subject and its applications, see [4, 7, 10, 13, 15, 28, 29]. For multivariate analysis and game problem, see [15, 17, 21, 27] and references cited therein.

In the case of the nonlinear complementarity problem, the associated information on the mapping F and the cone are implicitly assumed to be fixed and entirely independent from any other related parameters. However, this type of formulation is unable to model all realistic situations of the problem and fails to explain the complete reality. In the domains of engineering design and optimal control, for instance, [2], the problem's data involves a time parameter; in non-cooperative games, such as generalized Nash equilibrium [11], each player's strategy depends on the other players' strategies in case of the realistic model. Here we introduce semi-infinite tensor complementarity (SITCP), a generalized version of nonlinear complementarity to accommodate more number of realistic situations into the model.

The structure of the paper is as follows. Basic notations and findings are presented in Sect. 2 and are used in the next section. Under certain assumptions, we demonstrate in Sect. 3 that the solution set for the semi-infinite tensor complementarity problem exists. We develop the connection between the solution sets of the corresponding tensor complementarity problems and the semi-infinite tensor complementarity problem's solution set. In terms of the residual function, we determine the necessary and sufficient criteria for the error bound of the solution set to be level bounded.

2 Preliminaries

We begin by outlining the fundamental concepts and the notation that will be applied throughout the text. The vectors, matrices, and tensors are considered here with real entries. $[n]$ represents the set $\{1, ..., n\}$ for any positive integer n. Let $\mathbb{R}^n$ indicate the Euclidean space of dimension n and $\mathbb{R}^n_+$ indicate the set $\{x \in \mathbb{R}^n : x \geq 0\}$. Unless otherwise indicated, any real vector $x \in \mathbb{R}^n$ is a column vector. For the real vector $x \in \mathbb{R}^n$ define the Euclidean norm of x as $\|x\|_2 = \sqrt{|x_1^2| + \cdots |x_n^2|}$. The distance of x from $A \subseteq \mathbb{R}^n$ is denoted as $\text{dist}(x; A)$ and is defined as $\text{dist}(x; A) = \inf\{(x, a) : a \in A\}$. We denote the diameter of a set A as $\text{diam} A$ which is defined as $\text{diam} A = \sup_{x,y \in A} \|x - y\|$. $\mathbb{B}$ denote the unit ball in $\mathbb{R}^n$ such that $\mathbb{B} = \{x \in \mathbb{R}^n : \|x\| \leq 1\}$. The representation of a real tensor $\mathcal{A}$ with order m and dimension n is given by $\mathcal{A} = (a_{i_1 \ldots i_m})$, which is a multidimensional array with elements $a_{i_1 \ldots i_m} \in \mathbb{R}$ such that $i_j \in [n]$ and $j \in [m]$. $T_{m,n}$ represents the collection of all real tensor $\mathcal{A}$ with order m and dimension n. Shao [31] presented a tensor product that is used in the study of tensor complementarity theory. Let be two n-dimensional tensors, $\mathcal{A}$ of order $q \geq 2$ and $\mathcal{B}$ of order $k \geq 1$. The n dimension tensor $\mathcal{C}$ with order $(q - 1)(k - 1) + 1$ containing entries $c_{j\alpha_1 \cdots \alpha_{m-1}}$ is the product of $\mathcal{A}$ and $\mathcal{B}$. The entries of $\mathcal{C}$ are given by $c_{j\alpha_1 \cdots \alpha_{m-1}} = \sum_{j_2, \cdots, j_m \in [n]} a_{jj_2 \cdots j_m} b_{j_2\alpha_1} \cdots b_{j_m\alpha_{m-1}}$, where $j \in [n]$, $\alpha_1, \cdots, \alpha_{m-1} \in [n]^{k-1}$.

Given a real tensor $\mathcal{A} \in T_{m,n}$ and a real vector $x \in \mathbb{R}^n$, the quantity $\mathcal{A}x^{m-1} \in \mathbb{R}^n$ obtained by the Shao product is a real vector that is defined by

$$(\mathcal{A}x^{m-1})_i = \sum_{i_2,\ldots,i_m=1}^{n} a_{ii_2\ldots i_m} x_{i_2} \cdots x_{i_m}, \ \forall \ i \in [n],$$

and the quantity $\mathcal{A}x^m \in \mathbb{R}$ is a scalar that is defined by

$$\mathcal{A}x^m = x^T \mathcal{A}x^{m-1} = \sum_{i_1,\ldots,i_m=1}^{n} a_{i_1\ldots i_m} x_{i_1} \cdots x_{i_m}.$$

The set of feasible solutions of $\text{TCP}(q, \mathcal{A})$ is expressed as $\text{FEA}(q, \mathcal{A}) = \{x \in \mathbb{R}^n_+ : q + \mathcal{A}x^{m-1} \geq 0\}$, given a real vector $q \in \mathbb{R}^n$ and a real tensor $\mathcal{A} \in T_{m,n}$. The solution set of $\text{TCP}(q, \mathcal{A})$ is defined as $S = \text{SOL}(q, \mathcal{A}) = \{x \in \text{FEA}(q, \mathcal{A}) : x^T(q + \mathcal{A}x^{m-1}) = 0\}$. If $\exists$ any constant $c > 0$ (and $\epsilon > 0$) such that for each $x \in \mathbb{R}^n$ (where $r(x) \leq \epsilon$)

$$\text{dist}(x, S) \leq c \, r(x) \tag{3}$$

then the residual function $r(x)$ is said to be a global (local) error bound corresponding to the TCP. For the purposes of the upcoming parts, we take into consideration a few definitions and outcomes.

Definition 1 ([8]) For a tensor $\mathcal{A} = (a_{i_1 \dots i_m}) \in T_{m,n}$ the ith row subtensor is denoted as $R_i(\mathcal{A})$ with the elements given by $(R_i(\mathcal{A}))_{i_2 \dots i_m} = (a_{i i_2 \dots i_m})$, where $2 \leq j \leq m$ and $i_j \in [n]$.

Definition 2 ([35]) Consider the function $f : \mathbb{R}^n \to \mathbb{R}^n$. If $\forall \, \gamma \geq 0$ the level set of the function f, defined as $\{x \in \mathbb{R}^n : f(x) \leq \gamma\}$ is bounded then f is called level bounded .

Definition 3 ([32]) A tensor $\mathcal{A} \in T_{m,n}$ is called a S-tensor if the system

$$\mathcal{A}x^{m-1} > 0, \quad x > 0$$

has a solution.

Definition 4 ([32]) A tensor $\mathcal{A} \in T_{m,n}$ is called a P-tensor, if $\forall \, x \in \mathbb{R}^n \backslash \{0\}$, $\exists i \in [n]$ such that $x_i \neq 0$ and $x_i (\mathcal{A}x^{m-1})_i > 0$.

Definition 5 ([32, 34]) A tensor $\mathcal{A} \in T_{m,n}$ is called an R_0-tensor if zero is the unique solution of the TCP$(0, \mathcal{A})$.

Definition 6 ([37]) For a given $\epsilon \geq 0$, a residual function $r(x)$ of a TCP is said to be an ϵ-error bound if there exists $c > 0$ such that $\text{dist}(x, S) \leq c \, r(x) + \epsilon$ for all $x \in \mathbb{R}^n$. The concept reduces to the error bound if $\epsilon = 0$.

Theorem 1 ([22]) *If the intersection over any finite subcollection of A is non-empty, then the family A of non-empty subsets of $\mathbb{R}^n$ is said to have the finite intersection property (FIP).*

Theorem 2 ([1]) *For a P-tensor $\mathcal{A} \in T_{m,n}$ and any $q \in \mathbb{R}^n$ the solution set of TCP$(q, \mathcal{A})$ is non-empty and compact.*

Theorem 3 ([34]) *If $\mathcal{A} \in T_{m,n}$ is a R_0-tensor then the solution set of the TCP$(q, \mathcal{A})$ is bounded $\forall \, q \in \mathbb{R}^n$.*

3   Main Results

We begin with the introduction of semi-infinite tensor complementarity problem SITCP$(q(\omega), \mathcal{A}(\omega), \Omega)$.

The SITCP$(q(\omega), \mathcal{A}(\omega), \Omega)$ is to determine a real vector $x \in \mathbb{R}^n$ that satisfies

$$F(x, \omega) \geq 0, \quad x \geq 0 \quad \text{and } x^T F(x, \omega) = 0, \quad \text{forall } \omega \in \Omega \qquad (4)$$

where Ω is a set in $\mathbb{R}^p$ and $F : \mathbb{R}^n \times \Omega \to \mathbb{R}^n$ with $F = \mathcal{A}(\omega)x^{m-1} + q(\omega)$. The set $S^* = \text{SOL}(q(\omega), \mathcal{A}(\omega), \Omega) = \{x : \mathcal{A}(\omega)x^{m-1} + q(\omega) \geq 0, \, x \geq 0 \text{ and } x^T (\mathcal{A}(\omega)x^{m-1} + q(\omega)) = 0, \text{ forall } \omega \in \Omega\}$ represents the solution set of SITCP $(q(\omega), \mathcal{A}(\omega), \Omega)$.

Here the necessary and sufficient conditions for the non-emptiness of the set S^* are established.

Theorem 4 *Consider the SITCP$(q(\omega), A(\omega), \Omega)$. Let $A(\omega_0)$ be an R_0-tensor for some $\omega_0 \in \omega$. $S^* \neq \phi$ iff $\cap_{i=1}^{p} SOL(q(\omega_i), A(\omega_i)) \neq \phi$ for finite number of elements $\omega_1, \omega_2, ..., \omega_p \in \Omega$.*

Proof If part: Since $S^* \subseteq \cap_{i=1}^{p} SOL(q(\omega_i), A(\omega_i))$, $S^* \neq \phi$ implies that $\cap_{i=1}^{p} SOL(q(\omega_i), A(\omega_i)) \neq \phi$.

Only if part: Since $A(\omega_0)$ is an R_0-tensor, by Theorem 3, the set $SOL(q(\omega_0), A(\omega_0))$ is bounded. This further suggests that S^* is bounded. However, S^* is closed for every ω since $SOL(q(\omega), A(\omega))$ is closed. S^* is therefore compact. Hence the result from Theorem 1 of finite intersection of compact sets.

In the next result, we find the position of S^*. For this purpose, we consider the following:

(i) $(A_{\max})_{i_1 \cdots i_m} = (\bar{a}_{i_1 \cdots i_m}) = \max_{\omega \in \Omega} a_{i_1 \cdots i_m}(\omega)$
(ii) $(A_{\min})_{i_1 \cdots i_m} = (a'_{i_1 \cdots i_m}) = \min_{\omega \in \Omega} a_{i_1 \cdots i_m}(\omega)$
(iii) $(q_{\max})i = \max_{\omega \in \Omega} q_i(\omega)$
(iv) $(q_{\min})i = \min_{\omega \in \Omega} q_i(\omega)$

Theorem 5 *Consider the SITCP$(q(\omega), A(\omega), \Omega)$ with Ω being compact. If $q(\omega)$ and $A(\omega)$ are continuous on the compact set Ω, then $S^* \subseteq SOL(q_{\max}, M_{\max}) \cap SOL(q_{\min}, M_{\min})$. Furthermore, suppose in each row subtensor, $R_i(A(\omega))$ and in each row of $q(\omega)$, $q_i(\omega)$ the maximum (and minimum) is attained by a common $\bar{\omega}$ (and ω'), i.e., for each $i \in [n]$, there exist ω'_i, $\bar{\omega}_i \in \Omega$ such a way that $R_i(A_{\min}) = R_i(A(\omega'_i))$, $(q_{\min})_i = q(\omega'_i)_i$ and $R_i(A_{\max}) = R_i(A(\bar{\omega}_i))$, $(q_{\max})_i = q(\bar{\omega}_i)_i$. Then*

$$S^* = SOL(q_{\max}, A_{\max}) \cap SOL(q_{\min}, A_{\min}).$$

Proof By the rules of maximization and minimization for the summation of functions in Exercise 1.36 of [35], we obtain

$$\min_{\omega \in \Omega}(A(\omega)x^{m-1} + q(\omega)) \leq A_{\min}x^{m-1} + q_{\min} \tag{5}$$

$$\max_{\omega \in \Omega}(A(\omega)x^{m-1} + q(\omega)) \geq A_{\max}x^{m-1} + q_{\max} \tag{6}$$

for all $x \geq 0$. By using the inequalities (6) and (5) and using reasoning akin to that for Theorem 2.2 of [37], we have

$$S^* \subseteq SOL(q_{\max}, A_{\max}) \cap SOL(q_{\min}, A_{\min}). \tag{7}$$

Now to prove the second part let us assume that the conditions of Theorem 5 hold. Then $(A_{\min}x^{m-1})_i = (A(\omega')x^{m-1})_i$ and $(A_{\min}x^{m-1})_i + (q_{\min})_i = (A(\omega'_i)x^{m-1} + q(\omega_i))_i$. Therefore $SOL(q_{\min}, A_{\min}) = SOL(q(\omega'_i), A(\omega'_i))$. Similarly we have SOL $(q_{\max}, A_{\max}) = SOL(q(\bar{\omega}_i), A(\bar{\omega}_i))$. Since $S^* = \cap_{\omega \in \Omega} SOL(q(\omega), A(\omega))$, using (6) and (5) we conclude that

$$S^* = \text{SOL}(q_{\max}, \mathcal{A}_{\max}) \cap \text{SOL}(q_{\min}, \mathcal{A}_{\min}).$$

Here we provide an example to illustrate the result of Theorem 5.

Example 1 Let $\mathcal{A}(\omega) \in T_{3,2}$ and $q(\omega) \in \mathbb{R}^2$ be such that $a_{111} = (1 - 2\omega^3)$, $a_{121} = 1 - \omega$, $a_{112} = (1 - \omega)$, $a_{122} = -1$, $a_{211} = a_{212} = a_{221} = 0$ $a_{222} = -\omega^2$ and $q(\omega) = \begin{pmatrix} 1 \\ \omega^2 \end{pmatrix}$ and $\omega \in \Omega = [0, 1]$. Then $\mathcal{A}(\omega)x^2 = \begin{pmatrix} (1 - 2\omega^3)x_1^2 + 2(1 - \omega)x_1 x_2 - x_2^2 \\ -\omega^2 x_2^2 \end{pmatrix}$.
Now consider the SITCP($q(\omega)$, $\mathcal{A}(\omega)$, Ω) which is to determine a real vector $x = \begin{pmatrix} x_1 \\ x_2 \end{pmatrix} \in \mathbb{R}^2$ that satisfies the following requirements:

$$x_1 \geq 0; \quad (\mathcal{A}(\omega)x^2)_1 \geq 0; \quad x_1[(1 - 2\omega^3)x_1^2 + 2(1 - \omega)x_1 x_2 - x_2^2 + 1] = 0, \tag{8}$$

$$x_2 \geq 0; \quad (\mathcal{A}(\omega)x^2)_2 \geq 0; \quad x_2[-\omega^2 x_2^2 + \omega^2] = 0. \tag{9}$$

Solving Eqs. (8) and (9), we obtain the solution set for $\omega \in \Omega$, which is $\left\{ \begin{pmatrix} 0 \\ 0 \end{pmatrix}, \begin{pmatrix} 0 \\ 1 \end{pmatrix}, \begin{pmatrix} \frac{2(1-\omega)}{1-2\omega^3} \\ 1 \end{pmatrix} \right\}$. For $\omega = 1$ we get $\frac{2(1-\omega)}{1-2\omega^3} = 0$ so $S^* = \left\{ \begin{pmatrix} 0 \\ 0 \end{pmatrix}, \begin{pmatrix} 0 \\ 1 \end{pmatrix} \right\}$.
Now for the given tensor $\mathcal{A}(\omega)$, let $\mathcal{A}_{\max} = (\bar{a}_{ijk}) \in T_{3,2}$. Then we obtain $\bar{a}_{111} = 1$, $\bar{a}_{121} = \bar{a}_{112} = 1$, $\bar{a}_{122} = -1$ and $\bar{a}_{211} = \bar{a}_{212} = \bar{a}_{221} = \bar{a}_{222} = 0$ and $q_{max} = \begin{pmatrix} 1 \\ 1 \end{pmatrix}$.
Then the TCP($q_{\max}$, $\mathcal{A}_{\max}$) which is finding $x = \begin{pmatrix} x_1 \\ x_2 \end{pmatrix} \in \mathbb{R}^2$ such that

$$x_1 \geq 0; \quad (\mathcal{A}_{\max}x^2)_1 \geq 0; \quad x_1[x_1^2 + 2x_1 x_2 - x_2^2 + 1] = 0, \tag{10}$$

$$x_2 \geq 0; \quad (\mathcal{A}_{max}x^2)_2 \geq 0; \quad x_2[0 + 1] = 0. \tag{11}$$

Solving (10) and (11), we get $\text{SOL}(q_{\max}, \mathcal{A}_{\max}) = \left\{ \begin{pmatrix} 0 \\ 0 \end{pmatrix} \right\}$.
Again, let $\mathcal{A}_{\min} = (a'_{ijk}) \in T_{3,2}$. Then we obtain $a'_{111} = -1$, $a'_{121} = a'_{112} = 0$, $a'_{122} = -1$ and $a'_{211} = a'_{212} = a'_{221} = 0$, $a'_{222} = -1$ and $q_{min} = \begin{pmatrix} 1 \\ 0 \end{pmatrix}$. Then TCP($q_{\min}$, $\mathcal{A}_{\min}$) is finding $x = \begin{pmatrix} x_1 \\ x_2 \end{pmatrix} \in \mathbb{R}^2$ such that

$$x_1 \geq 0; \quad (\mathcal{A}_{\min}x^2)_1 \geq 0; \quad x_1[-x_1^2 + -x_2^2 + 1] = 0, \tag{12}$$

$$x_2 \geq 0; \quad (\mathcal{A}_{min}x^2)_2 \geq 0; \quad x_2[-x_2^2] = 0. \tag{13}$$

Solving (12) and (13), we have $\text{SOL}(q_{\min}, \mathcal{A}_{\min}) = \left\{ \begin{pmatrix} 0 \\ 0 \end{pmatrix}, \begin{pmatrix} 1 \\ 0 \end{pmatrix} \right\}$.

Thus $S^* \supset SOL(q_{max}, A_{max}) \cap SOL(q_{min}, A_{min})$, i.e., the inclusion is strict.

Now we replace $q(\omega)$ by $\bar{q} = \begin{pmatrix} 1 \\ 1 \end{pmatrix}$. Then the SITCP$(\bar{q}, A(\omega), \Omega)$ is to determine a

real vector $x = \begin{pmatrix} x_1 \\ x_2 \end{pmatrix} \in \mathbb{R}^2$ that satisfies the following requirements:

$$x_1 \geq 0; \quad (A(\omega)x^2)_1 \geq 0; \quad x_1[(1 - 2\omega^3)x_1^2 + 2(1 - \omega)x_1 x_2 - x_2^2 + 1] = 0,$$
$$(14)$$
$$x_2 \geq 0; \quad (A(\omega)x^2)_2 \geq 0; \quad x_2[-\omega^2 x_2^2 + 1] = 0. \tag{15}$$

Solving equations (14) and (15), we obtain the solution set for $\omega \in \Omega$, which is $\left\{ \begin{pmatrix} 0 \\ 0 \end{pmatrix}, \begin{pmatrix} \alpha(\omega) \\ \frac{1}{\omega} \end{pmatrix} \right\}$. Here $\alpha(\omega)$ is the positive root of the equation

$$\omega^2(1 - 2\omega^3)x_1^2 - 2\omega(1 - \omega)x_1 - (1 - \omega^2) = 0.$$

For $\omega = 1$ we get $\alpha(1) = 0$. Therefore $S^* = \left\{ \begin{pmatrix} 0 \\ 0 \end{pmatrix} \right\}$.

Then TCP$(q_{max}, A_{max}) = $ TCP$(\bar{q}, A_{max})$. SOL$(\bar{q}, A_{max}) = \left\{ \begin{pmatrix} 0 \\ 0 \end{pmatrix} \right\}$.

Now, TCP$(\bar{q}, A_{min})$ is finding $x = \begin{pmatrix} x_1 \\ x_2 \end{pmatrix} \in \mathbb{R}^2$ such that

$$x_1 \geq 0; \quad (A_{min}x^2)_1 \geq 0; \quad x_1[-x_1^2 + -x_2^2 + 1] = 0, \tag{16}$$

$$x_2 \geq 0; \quad (A_{min}x^2)_2 \geq 0; \quad x_2[-x_2^2 + 1] = 0. \tag{17}$$

Solving (16) and (17), we have SOL$(\bar{q}, A_{min}) = \left\{ \begin{pmatrix} 0 \\ 0 \end{pmatrix}, \begin{pmatrix} 1 \\ 0 \end{pmatrix}, \begin{pmatrix} 0 \\ 1 \end{pmatrix} \right\}$. In this case we have $S^* = SOL(q_{max}, A_{max}) \cap SOL(q_{min}, A_{min})$.

For the next result, we define semi-infinite S-tensor.

Definition 7 A tensor $A(\omega) \in T_{m,n}$ is called a semi-infinite S-tensor with respect to Ω if there exists a vector $x > 0$ satisfying $A(\omega)x^{m-1} > 0$, $\forall \omega \in \Omega$.

Here we establish a connection between the semi-infinite S-tensor and feasibility of the solution set of SITCP$(q(\omega), A(\omega), \Omega)$.

Theorem 6 *Consider an SITCP$(q(\omega), A(\omega), \Omega)$ where Ω is compact. Let all the elements of $A(\omega)$ are continuous on the set Ω. $A(\omega)$ is a semi-infinite S-tensor with respect to the set Ω iff the SITCP$(q(\omega), A(\omega), \Omega)$ is feasible $\forall q(\omega) \in C(\Omega)$, where $C(\Omega)$ is the set of all continuous function on Ω.*

Proof If part: Since $A(\omega)$ is a semi-infinite S-tensor, $\exists x > 0$ such that $A(\omega)x^{m-1} > 0$. Then $\exists$ a scalar $\lambda > 0$ (sufficiently small) such that $A(\omega)x^{m-1} \geq \lambda e > 0$, $\forall \omega \in$

Ω. Here $e = (1, 1, \cdots, 1)^T$. Now choose $\alpha > 0$ with $\alpha e > -q_{\min}$. Choosing $\bar{\alpha} = \left(\frac{\alpha}{\lambda}\right)^{\frac{1}{m-1}} > 0$ we have $\bar{\alpha}x > 0$. Also, $\mathcal{A}(\omega)(\bar{\alpha}x)^{m-1} = \frac{\alpha}{\lambda}\mathcal{A}(\omega)x^{m-1} \geq \alpha e > -q_{\min}$. Thus for $\bar{\alpha}x > 0$ we have $\mathcal{A}(\omega)(\bar{\alpha}x)^{m-1} + q(\omega) \geq \mathcal{A}(\omega)(\bar{\alpha}x)^{m-1} + q_{\min} > 0$. Therefore for SITCP$(q(\omega), \mathcal{A}(\omega), \Omega)$ the point $\bar{\alpha}x$ is feasible.

Only if part: Let for all $\omega \in \Omega$ $q(\omega) := \tilde{q} < 0$. Let the SITCP$(\tilde{q}, \mathcal{A}(\omega), \Omega)$ be feasible. Then $\exists$ a real vector $x \geq 0$ satisfying $\mathcal{A}(\omega)x^{m-1} + \tilde{q} \geq 0 \implies \mathcal{A}(\omega)x^{m-1} \geq -\tilde{q} > 0$ for all $\omega \in \Omega$. Since $\mathcal{A}(\omega)x^{m-1}$ is continuous on Ω there exists a sufficiently small $\lambda > 0$ such that $x + \lambda e > 0$, and $\mathcal{A}(\omega)(x + \lambda e)^{m-1} > 0$.

A necessary condition for the feasibility of the SITCP is given by the ensuing corollary.

Corollary 1 *Consider an SITCP$(q(\omega), \mathcal{A}(\omega), \Omega)$ where Ω is compact. Let $\mathcal{A}(\omega)$ is continuous on the set Ω. If $\mathcal{A}_{\min}$ is an S-tensor, then SITCP$(q(\omega), \mathcal{A}(\omega), \Omega)$ is feasible $\forall$ $q(\omega) \in C(\Omega)$.*

Proof Let $\mathcal{A}_{\min}$ be an S-tensor. Then for some $x > 0$ we have $\mathcal{A}_{\min}x^{m-1} > 0$. From the definition of $\mathcal{A}_{\min}$ it follows that $\mathcal{A}(\omega)x^{m-1} > 0 \forall \omega \in \Omega$. This implies that the tensor $\mathcal{A}(\omega)$ is a semi-infinite S-tensor. Hence using the Theorem 6, we conclude that SITCP$(q(\omega), \mathcal{A}(\omega), \Omega)$ is feasible $\forall$ $q(\omega) \in C(\Omega)$.

Now we establish an ϵ-error bound for the solution set of SITCP$(q(\omega), \mathcal{A}(\omega))$. Here ϵ represents the degree of approximation of the set S^*.

Theorem 7 *Consider an SITCP$(q(\omega), \mathcal{A}(\omega), \Omega)$. Let $S^* \neq \emptyset$. If for some $\omega_0 \in \Omega$, the tensor $\mathcal{A}(\omega_0)$ is an P-tensor then $\exists \epsilon > 0$ and $c > 0$ with $\epsilon \leq diam(SOL(q(\omega_0), \mathcal{A}(\omega_0))$ satisfying $dist(x, S^*) \leq c\ r(x) + \epsilon$ where residual function $r(x)$ is given by $r(x) = \max_{\omega\Omega} \| \min\{x, [\mathcal{A}(\omega)(x-y)]^{\frac{1}{m-1}} + [\mathcal{A}(\omega)y^{m-1}]^{\frac{1}{m-1}}\}\|$ and $y \in (SOL(q(\omega_0), \mathcal{A}(\omega_0))$.*

Proof Since $\mathcal{A}(\omega_0)$ is an P-tensor, SOL$(q(\omega_0), \mathcal{A}(\omega_0))$ is bounded. This implies $\exists$ an $\epsilon > 0$ such that SOL$(q(\omega_0), \mathcal{A}(\omega_0)) \subseteq S^* + \epsilon\mathbb{B}$. As a consequence, we obtain

$$dist(x, S^*) \leq dist(x, SOL(q(\omega_0), \mathcal{A}(\omega_0))) + \epsilon, \text{ for all } x \in \mathbb{R}^n. \tag{18}$$

Notice that $S^* \subseteq$ SOL$(q(\omega_0), \mathcal{A}(\omega_0))$. Therefore the diameter of SOL$(q(\omega_0), \mathcal{A}(\omega_0))$ provides an upper bound of ϵ. By the Theorem 3.2 of [36] for TCP$(q(\omega_0), \mathcal{A}(\omega_0))$, $\exists c > 0$ such that for all $x \in \mathbb{R}^n$,

$$dist(x, SOL(q(\omega_0), \mathcal{A}(\omega_0))) \leq c\| \min\{x, [\mathcal{A}(\omega_0)(x-y)]^{\frac{1}{m-1}} + [\mathcal{A}(\omega_0)y^{m-1}]^{\frac{1}{m-1}}\}\|. \tag{19}$$

From (18) and (19), the desired result follows.

For the next result, we define semi-infinite R_0-tensor and establish a connection between R_0-tensor and semi-infinite R_0-tensor.

Definition 8 The tensor $\mathcal{A}(\omega)$ is called a semi-infinite R_0-tensor with respect to Ω if zero is the unique solution of SITCP$(0, \mathcal{A}(\omega), \Omega)$, i.e.,

$$x \geq 0, \quad \mathcal{A}(\omega)x^{m-1} \geq 0, \quad x^T \mathcal{A}(\omega)x^{m-1} = 0, \quad \text{for all } \omega \in \Omega \implies x = 0. \quad (20)$$

Theorem 8 *Consider an SITCP$(q(\omega), \mathcal{A}(\omega), \Omega)$ with $\mathcal{A}(\omega_0)$ as an R_0-tensor for some $\omega_0 \in \Omega$. Then with respect to the set Ω the tensor $\mathcal{A}(\omega)$ is a semi-infinite R_0-tensor.*

Proof We have $\mathcal{A}(\omega_0)$ as an R_0-tensor. Then zero is the unique solution of TCP$(0, \mathcal{A}(\omega_0))$, i.e., SOL$(0, \mathcal{A}(\omega_0)) = \{0\}$. Also, $S^* \subseteq \cap_{\omega \in \Omega}$ SOL$(0, \mathcal{A}(\omega), \Omega)$ Therefore $S^* = \{0\}$. Hence the result.

Here, we demonstrate the necessary and sufficient criteria, expressed in terms of the residual function, for the error bound of the solution set to be level bounded. The semi-infinite tensor complementarity problem can be solved by identifying $x \in \mathbb{R}^n$ such that $\forall\, \omega \in \Omega, x \in$ SOL$(q(\omega), \mathcal{A}(\omega))$. For some ω, it is, nevertheless, frequently possible to produce $x \in$ SOL$(q(\omega), \mathcal{A}(\omega))$. In such a circumstance, it is crucial to give a numerical estimate of how close each $x \in \mathbb{R}^n$ is, in terms of certain residual functions $r(x)$, to each SOL$(q(\omega), \mathcal{A}(\omega))$. Stated otherwise, we identify $c > 0$ for which the following condition holds:

$$\text{dist}(x, \text{SOL}(q(\omega), \mathcal{A}(\omega))) \leq c\, r(x), \quad \forall\, \omega \in \Omega, \,\forall\, x \in \mathbb{R}^n$$

which is equivalent to

$$\max_{\omega \in \Omega} \text{dist}(x, \text{SOL}(q(\omega), \mathcal{A}(\omega))) \leq c\, r(x), \quad \forall\, x \in \mathbb{R}^n.$$

Here c is said to be a weak error bound. The significance of weak error bound is that the solution S^* does not need to be non-empty as required in case of error bound.

Theorem 9 *Consider an SITCP$(q(\omega), \mathcal{A}(\omega), \Omega)$ where Ω is compact. Let $q(\omega)$ and $\mathcal{A}(\omega)$ are continuous. Then the necessary and sufficient condition for the residual function, $r(x) = \max_{\omega \in \Omega} \| \min (x, \mathcal{A}(\omega)x^{m-1} + q(\omega))\|^2$ to be level bounded is that the tensor $\mathcal{A}(\omega)$ is a semi-infinite R_0-tensor with respect to Ω.*

Proof Only if part: We prove the first result by the contrapositive method. Using this method, we first presume that the residual function $r(x)$ is not level bounded. Therefore there exists a sequence $\{x_n\} \to \infty$ as $n \to \infty$, $\{r(x_n)\}$ is bounded. We assume that $\frac{x_n}{\|x_n\|^{m-1}}$ converge to the limit x_0 with $\|x_0\| = 1$. Using continuity of $q(\omega)$ and $\mathcal{A}(\omega)$ along with the compactness of the set Ω, we observe that the function $q(\omega)$ is bounded and $r(x)$ is continuous on the set Ω. Therefore, $\lim_{n \to \infty} \frac{r(x_n)}{\|x_n\|^{m-1}} = 0$ and $\lim_{n \to \infty} \frac{q(\omega)}{\|x_n\|^{m-1}} = 0$ for all $\omega \in \Omega$. Now

$$\frac{r(x_n)}{\|x_n\|^{2(m-1)}} = \max_{\omega \in \Omega} \| \min \left\{ \frac{x_n}{\|x_n\|^{m-1}}, \frac{\mathcal{A}(\omega)x + q(\omega)}{\|x_n\|^{m-1}} \right\} \|^2. \quad (21)$$

As $n \to \infty$, we find by taking the limit on both sides of the equation (21) that the following equation holds:

$$\max_{\omega \in \Omega} \| \min \{x_0, A(\omega)x_0^{m-1}\} \|^2 = 0.$$

This means that the SITCP$(0, A(\omega), \Omega)$ has a nonzero solution x_0. Hence $A(\omega)$ is not a semi-infinite R_0-tensor. This completes the proof.

If part: Assume that the SITCP$(0, A(\omega), \Omega)$ has a nonzero solution. Let x be a nonzero solution SITCP$(0, A(\omega), \Omega)$. Let $H(x) = \{i : x_i = 0\}$ and $G(x) = \{i : x_i > 0\}$. The continuity of $q(\omega)$ and compactness of Ω ensure that the function $q(\omega)$ is bounded on the set Ω. Therefore $\exists$ a scalar $\bar{d} > 0$ such that, for any $d > \bar{d}$

$$dx_i \geq q_i(\omega) \ \forall \ i \in G(x) \ \text{and} \ \forall \ \omega \in \Omega. \tag{22}$$

Given any $d > \bar{d}$ we have

$$r(dx) = \max_{\omega \in \Omega} \| \min(dx, d^{m-1}A(\omega)x^{m-1} + q(\omega)) \|^2 \tag{23}$$

$$\leq \sum_{i=1}^{n} \max_{\omega \in \Omega} \{\min(dx_i, (d^{m-1}A(\omega)x^{m-1})_i + q_i(\omega))\}^2.$$

Consider the following cases.
Case-1. First we consider the case when $i \in G(x)$. Then $((A(\omega)x^{m-1})_i = 0$. It follows from (22) that

$$\max_{\omega \in \Omega} \{\min\{dx_i, d^{m-1}(A(\omega)x^{m-1})_i + q_i(\omega))\}\}^2 = \max_{\omega \in \Omega} q_i(\omega)^2. \tag{24}$$

Case-2. We consider the case when $i \in H(x)$. If $d^{m-1}(A(\omega)x^{m-1})_i + q_i(\omega) \geq 0$, we have

$$\{\min(dx_i, d^{m-1}(A(\omega)x^{m-1})_i + q_i(\omega))\}^2 = 0. \tag{25}$$

If $d^{m-1}(A(\omega)x^{m-1})_i + q_i(\omega) < 0$, then by the fact $q_i(\omega) \leq d^{m-1}(A(\omega)x^{m-1})_i + q_i(\omega) < 0$ we obtain

$$\{\min(dx_i, d^{m-1}(A(\omega)x^{m-1})_i + q_i(\omega))\}^2 \leq q_i(\omega)^2. \tag{26}$$

Thus combining (25) and (26), we have

$$\max_{\omega \in \Omega} [\min(dx_i, d^{m-1}(A(\omega)x^{m-1})_i + q_i(\omega))]^2 \leq \max_{\omega \in \Omega} q_i(\omega)^2. \tag{27}$$

Putting the facts (23), (24), and (27) together, we find that

$$r(dx) \leq \sum_{i=1}^{n} \max_{\omega \in \Omega} q_i(\omega)^2 < \infty \quad \forall \, d \geq \bar{d}.$$

This contradicts the fact that $r(x)$ is level bounded.

4 Conclusion

In this article, we are introducing semi-infinite tensor complementarity problem to accommodate a more realistic situation of the problem. We show that the solution set of semi-infinite tensor complementarity problem exists with some assumption. An example is illustrated in detail to establish the result. We establish a connection between semi-infinite tensor and its equivalent tensors to obtain the solution of semi-infinite tensor complementarity problem. We show that the error bound of the solution set is level bounded in terms of residual function. Finally we propose for future research in the area of P-tensor, Z-tensor, and semi-positive tensor classes in the sense of semi-infinite tensor complementarity problem.

Acknowledgements The Council of Scientific and & Industrial Research (CSIR), India, is acknowledged by the author R. Deb for the financial support given by the Junior Research Fellowship programme.

References

1. Bai XL, Huang ZH, Wang Y (2016) Global uniqueness and solvability for tensor complementarity problems. J Optim Theory Appl 170(1):72–84
2. Chen Q, Chu D, Tan RC (2005) Optimal control of obstacle for quasi-linear elliptic variational bilateral problems. SIAM J Control Optim 44(3):1067–1080
3. Das AK (2016) Properties of some matrix classes based on principal pivot transform. Ann Oper Res 243(1):375–382
4. Das AK, Jana R (2016) Deepmala: on generalized positive subdefinite matrices and interior point algorithm. In: International conference on frontiers in optimization: theory and applications. Springer, pp 3–16
5. Das AK, Jana R (2017) Deepmala: finiteness of criss-cross method in complementarity problem. In: International conference on mathematics and computing. Springer, pp 170–180
6. Das AK, Jana R (2018) Deepmala: invex programming problems with equality and inequality constraints. Trans A. Razmadze Math Inst 172(3):361–371
7. Das AK, Jana R et al (2019) Some aspects on solving transportation problem. Yugoslav J Oper Res 30(1):45–57
8. Deb R, Das AK (2023) More on semipositive tensor and tensor complementarity problem. In: International conference on mathematics and computing. Springer, pp 147–156
9. Dutta A, Das AK (2023) On some properties of k-type block matrices in the context of complementarity problem. In: Mathematics and computing: ICMC 2022, Vellore, India, January 6–8. Springer, pp 143–154

10. Dutta A, Jana R, Das AK (2022) On column competent matrices and linear complementarity problem. In: Proceedings of the seventh international conference on mathematics and computing. Springer Singapore, pp 615–625
11. Facchinei F, Fischer A, Piccialli V (2009) Generalized nash equilibrium problems and newton methods. Math Program 117(1–2):163–194
12. Huang ZH, Qi L (2017) Formulating an n-person noncooperative game as a tensor complementarity problem. Comput Optim Appl 66(3):557–576
13. Jana R, Das AK, Dutta A (2019) On hidden Z-matrix and interior point algorithm. Opsearch 56(4):1108–1116
14. Jana R, Das AK, Mishra VN (2021) Iterative descent method for generalized leontief model. Proc Natl Acad Sci, India, Sect A 91(2):237–244
15. Jana R, Das AK, Sinha S (2018) On processability of Lemke's algorithm. Appl Appl Math 13(2)
16. Jana, R, Das, A.K, Sinha, S (2018) On semimonotone star matrices and linear complementarity problem. arXiv preprint arXiv:1808.00281
17. Jana R, Dutta A, Das AK (2021) More on hidden Z-matrices and linear complementarity problem. Linear Multilinear Algebra 69(6):1151–1160
18. Mohan SR, Neogy SK, Das AK (2001) More on positive subdefinite matrices and the linear complementarity problem. Linear Algebra Appl 338(1–3):275–285
19. Mohan SR, Neogy SK, Das AK (2001) On the classes of fully copositive and fully semimonotone matrices. Linear Algebra Appl 323(1–3):87–97
20. Mohan SR, Neogy SK, Das AK (2004) A note on linear complementarity problems and multiple objective programming. Math Program 100(2):339–344
21. Mondal P, Sinha S, Neogy SK, Das AK (2016) On discounted ARAT semi-markov games and its complementarity formulations. Internat J Game Theory 45(3):567–583
22. Munkres J (2016) Topology James Munkres, 2nd edn
23. Neogy SK, Bapat RB, Das AK (2016) Optimization models with economic and game theoretic applications. Ann Oper Res 243(1):1–3
24. Neogy SK, Das AK (2005) On almost type classes of matrices with Q-property. Linear and Multilinear Algebra 53(4):243–257
25. Neogy SK, Das AK (2005) Principal pivot transforms of some classes of matrices. Linear Algebra Appl 400:243–252
26. Neogy SK, Das AK (2006) Some properties of generalized positive subdefinite matrices. SIAM J Matrix Anal Appl 27(4):988–995
27. Neogy SK, Das AK (2008) Mathematical programming and game theory for decision making, vol 1. World Scientific
28. Neogy SK, Das AK (2011) On singular N_0-matrices and the class Q. Linear Algebra Appl 434(3):813–819
29. Neogy SK, Das AK, Bapat RB (2009) Modeling, computation and optimization, vol 6. World Scientific
30. Neogy SK, Das AK, Gupta A (2012) Generalized principal pivot transforms, complementarity theory and their applications in stochastic games. Optim Lett 6(2):339–356
31. Shao JY (2013) A general product of tensors with applications. Linear Algebra Appl 439(8):2350–2366
32. Song Y, Qi L (2015) Properties of some classes of structured tensors. J Optim Theory Appl 165(3):854–873
33. Song Y, Qi L (2017) Properties of tensor complementarity problem and some classes of structured tensors. Ann Appl Math
34. Song Y, Yu G (2016) Properties of solution set of tensor complementarity problem. J Optim Theory Appl 170(1):85–96
35. Wets RRRB (1998) Variational analysis. Springer, Berlin

36. Zheng M, Zhang Y, Huang ZH (2019) Global error bounds for the tensor complementarity problem with a p-tensor. J Indust Manag Optim 15(2):933
37. Zhou J, Xiu N, Chen JS (2013) Solution properties and error bounds for semi-infinite complementarity problems. J Indust Manag Optim 9(1):99–115

Geometric Algorithm for Generalized Inverse of Rank Deficient Real Matrices

Phani Kumar Nyshadham, Levin Dabhi, Archie Mittal, and Harsh Kedia

Abstract An inverse of rank deficient matrices (*Generalized Inverse*) is applied to solve ill-conditioned problems such as large-sized matrix computations. Generalized inverses have many applications in engineering problems, such as data analysis, electrical networks, character recognition, and so on. The most frequently used Moore-Penrose inverse matrices allow for solving such systems, even with rank deficiency, and they provide minimum-norm vectors as solutions. In this paper, we propose novel geometric algorithm for computing generalized inverse of rank deficient real matrices. While some of the approaches for the formulations are purely based on *LU*-factorization, the other variations are based on *LU* and *QR* factorizations. The uniqueness of the generalized inverse is also proved for the proposed formulations.

Keywords SVD · Pseudo inverse · Matrix rank · Rank deficiency

1 Introduction

Generalized inverses are a powerful tool in linear algebra that allow us to find a solution to a wide range of problems that do not have a unique solution using traditional matrix inversion. In many practical applications, we often encounter situations where the matrix is singular, ill-conditioned, or not square, making it impossible to find a unique solution. In such cases, they provide a way to overcome these limitations and obtain a solution that best approximates the desired outcome. In fields such as engineering, physics, economics, and statistics, they are used to solve linear equations, linear least squares problems, etc. In deep learning, they are used for matrix factor-

P. K. Nyshadham (✉) · L. Dabhi · A. Mittal · H. Kedia
Intel Corporation, Bangalore, India
e-mail: phani.kumar.nyshadham@intel.com

L. Dabhi
e-mail: levin.dabhi@intel.com

A. Mittal
e-mail: archie.mittal@intel.com

ization where they aid in finding a low-rank approximation of a high-dimensional matrix. They are used in linear regression, where they find a solution that balances the goodness of fit and the complexity of the model.

Computing generalized inverses is an important problem in linear algebra, and there are many techniques available to solve it. One of the most common techniques for computing generalized inverses is the Moore-Penrose inverse [1], which is also known as the pseudo inverse. The Moore-Penrose inverse is defined for any matrix, regardless of its properties, and is unique. Other techniques for computing generalized inverses include the Drazin inverse [2], which is useful for matrices that have singular submatrices, and the group inverse, which is useful for matrices that have a certain group structure. These techniques are specialized and are used in specific applications. There are also iterative methods for computing generalized inverses, such as the iterative refinement method and the Newton method [3]. These methods can be useful when the matrix is large and sparse, and the computation of the generalized inverse is time-consuming.

Overall, the choice of technique for computing a generalized inverse depends on the properties of the matrix and the problem at hand. Linear algebra provides a powerful framework for solving problems in diverse fields, and the computation of generalized inverses is a crucial part of this framework.

2   Existing Algorithms for Generalized Inverses

In this section, we briefly describe a few existing algorithms for computing the generalized inverse.

2.1   The Moore-Penrose Inverse [5]

The Moore-Penrose inverse allows for solving least square systems, even with rank deficient matrices, in such a way that each column vector of the solution has a minimum norm. The Moore-Penrose inverse of a $m \times n$ matrix G is the unique $n \times m$ matrix G^+. Whenever G is of full rank n, the Moore-Penrose inverse reduces to the usual pseudo inverse $G^+ = (G^T * G)^{-1} G$. However, when G is rank deficient (that is when its rank $r < \min(m, n)$), the computation of G^+ is more complex.

C. C. MacDuffee apparently was the first to point out that a full rank factorization of a matrix A leads to an explicit formula for its Moore-Penrose inverse, $A^\dagger$, explained in the following theorem.

Theorem: If $A \in C_r^{m \times n}, r > 0$, has a full rank factorization $A = F * G$, then $A^\dagger = G^* (F^* A G^*)^{-1} F^*$.

There are several methods for computing Moore-Penrose inverse matrices [5]. The most commonly used is the Singular Value Decomposition (SVD) method that is implemented, for example, in the "pinv" function of Matlab (version 6.5.1), as well as in the "PseudoInverse" function of Mathematica (version 5.1). The Moore-Penrose inverse of A, denoted by A^+, by SVD method is given by $A^+ = V \Sigma^+ U^T$, where Σ^+ is a diagonal matrix containing the reciprocals of the nonzero singular values of A, and the superscript $^+$ denotes the transpose and conjugate of a matrix. This method is very accurate, but time-consuming when the matrix is large.

2.2 Fast Computation of Moore-Penrose Inverse Matrices [1]

Consider the symmetric positive $n \times n$ matrix $G'G$, and assume that is of rank $r \leq n$ and if there exists a unique upper triangular matrix S with exactly $n - r$ zero rows, and such that $S'S = G'G$, while the computation of S is a simple extension of the usual Cholesky factorization of non-singular matrices. Removing the zero rows from S, one obtains a $r \times n$ matrix, say L', of full rank r, and the formulation of inverse G^+ is obtained through the following steps:

$$G'G = S' \tag{1}$$
$$S = LL' \tag{2}$$
$$G^+ = L \left(L'L\right)^{-1} \left(L'L\right)^{-1} L'G'$$

There have been iterative algorithms to compute Moore-Penrose inverse matrices in the literature [5].

3 Proposed Algorithm for Generalized Inverse for Rank Deficient Matrices

In this section, we propose novel geometric algorithm for computing the generalized inverse of rank deficient real matrices.

For the sake of discussion, we start with a matrix A of order $m \times n$ (m rows, n columns) with rank r where $r \leq \min(m, n)$ implying rank deficiency. We impose some requirements on A which can be seen as applying permutations on the rows (P_1) and columns (P_2) of the original matrix A_{orig} to obtain A:

$$A = P_1 * A_{\text{orig}} * P_2 \tag{3}$$

For the system of equations defined by $A_{\text{orig}} * x_{\text{orig}} = b_{\text{orig}}$, the permuted system would be

$$A * x = b \quad \text{where}$$
$$A = P_1 * A_{\text{orig}}$$
$$b = P_1 * b_{\text{orig}}$$
$$x = P_2 * x_{\text{orig}} \tag{4}$$

These requirements on A are needed to simplify the foregoing algorithmic formulation for generalized inverse by ensuring the following conditions:

1. P_1 is chosen such that the first r rows of A are linearly independent

2. P_2 is chosen such that the first r columns of A are linearly independent.

In order to find the Least Squares (LS) solution $\vec{x}$ to the system $A\vec{x} = \vec{b}$ [4] for such matrix A, the first step is to find the orthogonal projection of $\vec{b}$ onto the $COLUMN\text{-}SPACE$ of A denoted by $R(A)$. To find the orthogonal projection of $\vec{b}$ onto $R(A)$, we consider a matrix $C_{m \times m}$ to be constructed from the basis vectors of $R(A)$ and also from the basis vectors of $\ker(A^T)$ referred to as the $LEFT\text{-}NULLSPACE$ of A where $\ker(A^T) \perp R(A)$ as described below

$$C = [\text{Bases}_{R(A)} | \text{Bases}_{\ker(A^T)}] \tag{5}$$

It is to be noted that C is full rank matrix and hence non-singular. We proceed to find the solution $\vec{z}$ to the below linear system where $\vec{b} \in R(C)$ and $R(C) = \Re^m$ denotes the column space of C:

$$C\vec{z} = \vec{b}$$
$$\vec{z} = C^{-1}\vec{b} \tag{6}$$

To find the orthogonal projection $p(\vec{b})$ onto $R(A)$, we define Δ_3 matrix of order $m \times m$ such that the orthogonal projection of $\vec{b}$ $(p(\vec{b}))$ onto $R(A)$ is defined as follows:

$$p(\vec{b}) = C * \Delta_3 * \vec{z}$$
$$p(\vec{b}) = C * \Delta_3 * C^{-1}\vec{b} \tag{7}$$

where Δ_3 is defined as

$$\Delta_3(i, j) = \begin{cases} 1, & i = j \text{ and } 1 \leq i \leq r \\ 0, & \text{otherwise} \end{cases} \tag{8}$$

Since the first r columns of C are the bases of $R(A)$, which are same as the first r columns of A, we define Δ_2 matrix of order $n \times m$ such that $p(\vec{b})$ onto $R(A)$ can be reformulated as

$$p(\vec{b}) = A * \Delta_2 * \vec{z}$$
$$p(\vec{b}) = A * \Delta_2 * C^{-1}\vec{b} \qquad (9)$$

where Δ_2 is defined as

$$\Delta_2(i,j) = 1 \begin{cases} 1, & i = j \text{ and } 1 \leq i \leq r \\ 0, & \text{otherwise} \end{cases} \qquad (10)$$

As we obtained the orthogonal projection of $\vec{b}$ onto $R(A)$, the *LS* problem is reformulated based on the above equations as follows:

$$A\vec{x} = \vec{b} \Leftrightarrow A\vec{x} = p(\vec{b}) \qquad (11)$$
$$A\vec{x} = A * \Delta_2 * C^{-1}\vec{b} \qquad (12)$$
$$\vec{x} = \Delta_2 * C^{-1}\vec{b} \qquad (13)$$

From theory, because of rank deficiency of A, there may not be unique solution for the system $A\vec{x} = p(\vec{b})$. From the property of generalized inverse [5], generalized inverse always tries to find *MINIMUM-NORM* solution to the system of Eq. (11) as described above. In order to achieve a *minimum-norm* solution we propose to construct a new matrix B of order $n \times n$ from the basis vectors of $R(A^T)$, referred to as a $ROW - SPACE$ of A, along with the basis vectors of $\ker(A)$, referred to as the $NULLSPACE$ of A, where $\ker(A) \perp R(A^T)$ is as described below

$$B = [\text{Bases}_{R(A^T)} | \text{Bases}_{\ker(A)}] \qquad (14)$$

As $R(B) = \Re^n$, the *LS* solution $\vec{x}$ is spanned by the columns of B as defined below

$$\vec{x} = B * \vec{y} \qquad (15)$$
$$B * \vec{y} = \Delta_2 * C^{-1}\vec{b}; \qquad (\because Eq.(13))$$
$$\vec{y} = B^{-1} * \Delta_2 * C^{-1}\vec{b} \qquad (16)$$

To interpret the above equations geometrically, we are trying to find the spanning coefficients of the columns of B represented by $\vec{y}$ to span one of the possibly infinitely many solutions given by $\vec{x} = \Delta_2 * C^{-1}\vec{b}$ which solve $A\vec{x} = p(\vec{b})$. It is to be noted that B is full rank and hence a non-singular matrix whose first r column vectors span the row-space of A denoted by $R(A^T)$. The last $n - r$ columns of B span the Nullspace of A denoted by $\ker(A)$. We define one more matrix Δ_1 of order $n \times n$ to set the last $n - r$ elements of $\vec{y}$ to $0's$ to produce $\vec{y}_r$ as follows:

$$\vec{y}_r = \Delta_1 * \vec{y} \qquad (17)$$

where Δ_1 is defined as

$$\Delta_1(i,j) = \begin{cases} 1, & i = j \text{ and } 1 \le i \le r \\ 0, & \text{otherwise} \end{cases} \tag{18}$$

By multiplying back B with $\vec{y}_r$, we produce only the $row - space$ component of the LS solution vector $\vec{x} = \Delta_2 * C^{-1}\vec{b}$ denoted by

$$\vec{x}_{LS,\text{min}_{\text{norm}}} = B * \vec{y}_r \tag{19}$$

$$\vec{x}_{LS,\text{min}_{\text{norm}}} = B * \Delta_1 * \vec{y}$$

$$\vec{x}_{LS,\text{min}_{\text{norm}}} = B * \Delta_1 * B^{-1} * \Delta_2 * C^{-1}\vec{b} \tag{20}$$

Since it is a known fact that the $row - space$ component of the infinitely many solutions of $A\vec{x} = p(\vec{b})$ is $UNIQUE$ and it is also the $MINIMUM$ -$NORM$ solution, the generalized inverse of A (A^+) is defined as

$$A^+ = B * \Delta_1 * B^{-1} * \Delta_2 * C^{-1} \tag{21}$$

With the above derivation, the generalized inverse of A_{orig} and the LS-$MINIMUM$ -$NORM$ solution x_{orig} of $A_{\text{orig}} * x_{\text{orig}} = b_{\text{orig}}$ are given by

$$A_{\text{orig}}^+ = B * \Delta_1 * B^{-1} * \Delta_2 * C^{-1} * P_1$$

$$x_{\text{orig}} = P_2^{-1} * B * \Delta_1 * B^{-1} * \Delta_2 * C^{-1} * P_1 * \vec{b}_{\text{orig}}$$

3.1 Proof of Uniqueness

As described in the above section, the computation of generalized inverse depends on the construction of C and B matrices. For the construction of matrix C, one may even choose to produce orthogonal bases only for the $ker(A^T)$ by following Gram-Schmidt orthogonalization or other techniques. It is to be noted that the first r columns of C which are the basis vectors of $R(A)$ are the first r columns of A - which means the construction of C is not UNIQUE, because to find an orthogonal projection of $\vec{b}$ onto the $R(A)$, we can form C with any bases of $ker(A^T)$. By zeroing out the $LEFT$-$NULLSPACE$ components and retaining only the $COLUMN - SPACE$ components, we find the orthogonal projection of $\vec{b}$ onto $R(A)$. Hence, the projection vector $(p(\vec{b}))$ is UNIQUE for a given $\vec{b}$, irrespective of the bases we choose for constructing C (albeit we choose a different set of orthogonal/non-orthogonal bases for not only $ker(A^T)$ but also for $R(A)$).

Assume we have C_1 also such that

$$p(\vec{b}) = C * \Delta_3 * C^{-1}\vec{b} = A * \Delta_2 * C^{-1}\vec{b}$$

$$p(\vec{b}) = C_1 * \Delta_3 * C_1^{-1}\vec{b} = A * \Delta_2 * C_1^{-1}\vec{b}$$

which implies

$$C * \Delta_3 * C^{-1}\vec{b} = C * \Delta_3 * C_1^{-1}\vec{b} \quad \because C_{(m \times r)} = C_{1(m \times r)}$$

$$(\Delta_3 * C^{-1} - \Delta_3 * C_1^{-1}) * \vec{b} = \vec{0}$$

$$E * \vec{b} = \vec{0}; \quad \text{where} \quad E = (\Delta_3 * C^{-1} - \Delta_3 * C_1^{-1}) \quad (22)$$

The above equation should be satisfied for every $\vec{b}$ implying the whole space R^m of all possible $\vec{b}$ is the NULL-SPACE of matrix E which is possible only when $E = 0$ implies $\Delta_3 * C^{-1} = \Delta_3 * C_1^{-1}$.

Similarly considering the row-space of A ($R(A^T)$) and null-space of A (ker(A)), one can choose any bases respectively to form matrix B. Which means, from the geometrical interpretation, for a given orthogonal projection vector $p(\vec{b})$ of a given $\vec{b}$ onto $R(A)$, we can consider any solution $\vec{x}_1$ (out of infinite solutions) for $A\vec{x} = p(\vec{b})$ such that $A\vec{x}_1 = p(\vec{b})$. With B matrix constructed as described in the above section, as repeated below

$$B * \vec{y} = \vec{x}_1 \implies \vec{y} = B^{-1}\vec{x}_1$$

$$\vec{x}_{1,\text{rowspace}} = B * \Delta_1 * \vec{y} = B * \Delta_1 * B^{-1}\vec{x}_1 \quad (23)$$

If we consider the same solution $\vec{x}_1$ (out of infinite solutions) for $A\vec{x} = p(\vec{b})$ such that $A\vec{x}_1 = p(\vec{b})$, assume that B_1 is also constructed on similar lines of B by choosing different bases for $R(A^T)$ and ker(A),

$$B_1 * \vec{z} = \vec{x}_1 \implies \vec{z} = B_1^{-1}\vec{x}_1$$

$$\vec{x}_{2,\text{rowspace}} = B_1 * \Delta_1 * \vec{z} = B_1 * \Delta_1 * B_1^{-1}\vec{x}_1 \quad (24)$$

From theory, the row-space vector $\vec{x}_r$ mapping to a column space vector $\vec{p}$ such that $A\vec{x}_r = \vec{p}$ is UNIQUE. From the above equations, for the same column space vector $p(\vec{b})$, $\vec{x}_{1,\text{rowspace}} = \vec{x}_{2,\text{rowspace}}$, which implies

$$B * \Delta_1 * B^{-1}\vec{x}_1 = B_1 * \Delta_1 * B_1^{-1}\vec{x}_1$$

$$(B * \Delta_1 * B^{-1} - B_1 * \Delta_1 * B_1^{-1})\vec{x}_1 = \vec{0}$$

$$F\vec{x}_1 = \vec{0}; \quad \text{where} \quad F = (B * \Delta_1 * B^{-1} - B_1 * \Delta_1 * B_1^{-1})$$
$$(25)$$

The above equation should be satisfied for $\vec{x}_1 = p(\vec{b}) \in R(A)$ arising from orthogonal projection of every $\vec{b}$. We can say for different choices of $\vec{b}$, the whole $R(A)$ is spanned by the corresponding orthogonal projections, which means for every vector $\vec{c} \in R(A)$, if we consider any $\vec{x}_1$ as the solution vector such that $A\vec{x}_1 = \vec{c}$, which means the space of $\vec{x}_1$ is R^n. From the above equations, it implies that the whole space R^n of all possible $\vec{x}_1$ is the NULL-SPACE of matrix F which is possible only when $F = 0$ implies $B * \Delta_1 * B^{-1} = B_1 * \Delta_1 * B_1^{-1}$.

Hence the generalized inverse A^+ as defined below is proved to be UNIQUE

$$A^+ = B * \Delta_1 * B^{-1} * \Delta_2 * C^{-1} \tag{26}$$

3.2 Construction of Generalized/Pseudo Inverse Based on LU Factorization

In this section, we present the scheme of construction of the crucial matrices C and B for computing the pseudo inverse as discussed in the previous section. Firstly from the previous discussion, C is constructed as

$$C = [\text{Basis}_{R(A)} | \text{Basis}_{\ker(A^T)}]$$

The construction of $R(A)$ comprises finding the linearly independent columns of A matrix. The linearly independent columns of A are identified by performing LU decomposition of A as $PA = LU$. Secondly, for finding the $ker(A^T)$ which is referred to as the left-nullspace of A, we compute the linear combinations of the linearly independent rows of A producing the other dependent rows of A. Each of the combinations forms a column inside $ker(A^T)$. Similarly in the second part, we proceed to construct B as

$$B = [\text{Basis}_{R(A^T)} | \text{Basis}_{\ker(A)}]$$

As described above $R(A^T)$ comprises finding the linearly independent rows of U matrix. Transposing every such linearly independent row of U matrix into a column forms $\text{Basis}_{R(A^T)}$. As the last step, for finding the $\ker(A)$ which is referred to as the Nullspace of A, we solve the linear system $Ux = 0$ and every solution x forms a Basis vector for $\ker(A)$. Thus $\text{Basis}_{\ker(A)}$ is constructed by the solution vectors of $Ux = 0$.

4 Accuracy Evaluation—Simulation Results

In this section, we compare the accuracy of the proposed algorithm *GenPinv* implemented using LU decomposition against the SVD-based Matlab's *pinv* function. Both the algorithms were implemented on Matlab Release 2016b on 11th Gen Intel(R) Core(TM) i7 @ 3.00 GHz. Double point precision was chosen so as to achieve higher accuracy in the computed matrix. We divided our experiments into 3 sets. For all the 3 sets of experiments, input matrix was a rank deficient matrix and the values were generated randomly in the range of [−1, 1] (see Table 1).

Table 1 Accuracy comparison—proposed method versus MATLAB's pinv method

S. no.	Number of rows	Number of columns	Rank	Abs(pinv–GenPinv)
1	1012	1024	1000	5.0563e–09
2	1012	2048	1000	1.6399e–08
3	1012	3512	1000	4.6609e–10
4	1012	4512	1000	1.5466e–10
5	1012	5012	1000	5.0128e–11
6	1024	1012	1000	6.1557e–09
7	2048	1012	1000	6.6894e–08
8	3512	1012	1000	5.9224e–09
9	4512	1012	1000	6.8478e–10
10	5012	1012	1000	3.4705e–09
11	5512	5512	1000	6.0711e–12
12	5512	5512	1500	2.4839e–11
13	5512	5512	2000	1.6055e–09
14	5512	5512	2500	2.9074e–07
15	5512	5512	3000	2.8760e–06

In the first part of the table (S. no: 1–5), we fixed the number of rows and the rank of the matrix and varied the number of columns. With this setting, the maximum absolute difference between our method and SVD decreased with increasing number of columns. In the second part of the table (S. no: 6–10), we fixed the columns and rank while varying the number of rows of the matrix. Similar kind of results were seen with this setting as well.

In the last part of the table (S. no: 11–15), we fixed the rows and columns and varied the rank of matrix. With this setting it is observed that as we moved from more rank deficient toward less rank deficient matrices, the maximum absolute difference between the two algorithms increased because of the rounding errors becoming more prominent.

5 Other Variations of the Proposed Generalized Inverse

We propose some more variations to the proposed algorithm in the previous section for some special cases of matrices. We also show equivalence relationships between different formulations by performing QR decomposition in some cases and prove the variation formulae are equivalent to the formula derived in Sect. 3. As mentioned in Sect. 3, after constructing C matrix, we proceed with QR factorization of the matrix $C = [R(A)|N(A^T)]$ as $C = Q_c * R_c$, where Q_c is orthogonal and R_c is the upper-triangular matrix.

5.1   Full Column Rank Matrices

A matrix is said to be having full column rank if the rank (r) of $m \times n$ matrix A is same as the number of columns $(r = n$ and $m > n)$. Considering Eq. (11), with the QR decomposition as described above, the orthogonal projection of $\vec{b}$ can also be expressed in terms of the orthogonal matrix Q_c as follows:

$$p(\vec{b}) = Q_c * \Delta_3 * Q_c^{-1}\vec{b}$$

With the new formulation of $p(\vec{b})$ in terms of Q_c, the LS problem can be expressed as

$$A\vec{x} = Q_c * \Delta_3 * Q_c^{-1}\vec{b}$$

As $C = [A|N(A^T)]$ is factorized into product of Q_c and R_c, the above equation can be expressed as

$$A\vec{x} = C\vec{x}_{\text{ext}}$$
$$Q_c * R_c * \vec{x}_{\text{ext}} = Q_c * \Delta_3 * Q_c^{-1}\vec{b} \tag{27}$$

where $\vec{x}_{\text{ext}}$ is an extended vector whose last $m - r$ components are zeros as defined as $x_{\text{ext}}^T = [x^T |\text{zeros}_{m-r\times 1}^T]$. The above equation can be simplified as

$$R_c * \vec{x}_{\text{ext}} = \Delta_3 * Q_c^{-1}\vec{b}$$
$$\vec{x}_{\text{ext}} = R_c^{-1} * \Delta_3 * Q_c^{-1}\vec{b} \tag{28}$$

Because $r = n$, x is the unique LS solution of $Ax = b$ and is given by the first $r = n$ components of x_{ext}.

Hence for this case of full column rank matrix A, the pseudo inverse is given by

$$A^+ = R_c^{-1} * \Delta_3 * Q_c^{-1} \tag{29}$$

As per Eq. (21), the pseudo inverse of a matrix $A_{m\times n}$ is given by

$$A^+ = B * \Delta_1 * B^{-1} * \Delta_2 * C^{-1}$$

Since A is having full column rank, Δ_1 is reduced to IDENTITY matrix of order $n \times n$. Hence the pseudo inverse for this special case of A is given by

$$A^+ = \Delta_2 * C^{-1}$$
$$A^+ = \Delta_2 * R_c^{-1} * Q_c^{-1} \tag{30}$$

Equations (29) and (30) compute the same pseudo inverse of A except that Eq. (29) computed the extended solution vector whose first $r = n$ components are the actual solution, whereas the latter computes the actual solution vector precisely.

5.2 Full Row Rank Matrices

A matrix is said to be having full row rank if the rank (r) of $m \times n$ matrix A is same as the number of rows $(r = m$ and $m < n)$. We can see that $C = [R(A)]$, since the matrix is having full row rank. The LS problem can be expressed as $A\vec{x} = b$. As the first $r = m$ columns of A are same as that of C, A can be written as $= A = [C|D]$, and the LS system considering QR decomposition of C can be expressed as

$$[C|D]\vec{x} = \vec{b}$$
$$[Q_c * R_c|D] * \vec{x} = Q_c * Q_c^{-1}\vec{b} \tag{31}$$

If we set the $n - r$ last components of the $\vec{x}$ to zero, the above system becomes

$$Q_c * R_c * \vec{x_{\text{trim}}} = Q_c * Q_c^{-1}\vec{b}$$
$$\vec{x_{\text{trim}}} = R_c^{-1} * Q_c^{-1}\vec{b}$$
$$x^T = [x_{\text{trim}}^T|\text{zeros}_{n-r\times1}^T] \tag{32}$$

Because $r = m$, x is not the unique LS solution of $Ax = b$. To find the minimum norm solution $x_{\text{min}_{\text{norm}}}$, B matrix is constructed as expressed in Eq. (14) and x is expressed in terms of B as per Eq. (15) as $\vec{x} = B * \vec{y}$. The above equation can be rewritten as

$$B * \vec{y} = \begin{bmatrix} R_c^{-1} * Q_c^{-1}\vec{b} \\ \text{zeros}_{n-r\times1} \end{bmatrix} \tag{33}$$
$$\vec{y} = B^{-1} * \begin{bmatrix} R_c^{-1} * Q_c^{-1}\vec{b} \\ \text{zeros}_{n-r\times1} \end{bmatrix} \tag{34}$$

With Δ_1 as defined in the last section, the minimum norm solution $x_{\text{min}_{\text{norm}}}$ is written as

$$\vec{x_{\min_{\text{norm}}}} = B * \Delta_1 * B^{-1} * \begin{bmatrix} R_c^{-1} * Q_c^{-1}\vec{b} \\ \text{zeros}_{n-r\times 1} \end{bmatrix} \quad (35)$$

Hence for this case of full row rank matrix A, the pseudo inverse is given by

$$A^+ = B * \Delta_1 * B^{-1} * \begin{bmatrix} R_c^{-1} * Q_c^{-1} \\ \text{zeros}_{n-r\times m} \end{bmatrix} \quad (36)$$

$$A^+ = B * \Delta_1 * (B^{-1})_{n\times r} * R_c^{-1} * Q_c^{-1} \quad (37)$$

As per Eq. (21), the pseudo inverse of a matrix $A_{m\times n}$ is given by

$$A^+ = B * \Delta_1 * B^{-1} * \Delta_2 * C^{-1}$$

From Eq. (36), we can see that

$$\Delta_2 * C^{-1} = \begin{bmatrix} R_c^{-1} * Q_c^{-1} \\ \text{zeros}_{n-r\times m} \end{bmatrix} \quad (38)$$

$$\implies B^{-1} * \Delta_2 * C^{-1} = (B^{-1})_{n\times r} * R_c^{-1} * Q_c^{-1} \quad (39)$$

which implies that Eqs. (37) and (21) are the same.

5.3 Any Rank Deficient Matrix $A_{m\times n}$ with Rank 'r' Where $r < \min(m, n)$

For a given rank deficient matrix $A_{m\times n}$ with rank 'r' where $r < \min(m, n)$, we construct $C = [R(A)|N(A^T)]$ matrix; the orthogonal projection as per Eq. (11) considering QR decomposition of C can be written as below

$$p(\vec{b}) = C * \Delta_3 * C^{-1}\vec{b}$$
$$p(\vec{b}) = Q_c * R_c * \Delta_3 * R_c^{-1} * Q_c^{-1}\vec{b} \quad (40)$$

As the first $r < \min(m, n)$ columns of A are same as that of C, $A_{\text{ext}} = [A_{m\times r}| N(A^T)|A_{m\times n-r}]$ can be written as $= A_{\text{ext}} = [C|D]$, and the LS system can be expressed as

$$Ax = p(\vec{b}) \implies A_{\text{ext}}x_{\text{ext}} = p(\vec{b}); \qquad x_{\text{ext}} = \begin{bmatrix} x_{r\times 1} \\ \text{zeros}_{m\times 1} \\ x_{n-r\times 1} \end{bmatrix} \quad (41)$$

$$[C|D]x_{\text{ext}} = p(\vec{b})$$
$$[Q_c * R_c|D] * x_{\text{ext}} = Q_c * R_c * \Delta_3 * R_c^{-1} * Q_c^{-1}\vec{b} \quad (42)$$

If we set the $n - r$ last components of the $\vec{x}_{ext}$ to zero, the above system becomes

$$Q_c * R_c * \vec{x}_{trim} = Q_c * R_c * \Delta_3 * R_c^{-1} * Q_c^{-1} \vec{b}$$
$$\vec{x}_{trim} = \Delta_3 * R_c^{-1} * Q_c^{-1} \vec{b}$$
$$x_{ext}^T = [x_{trim}^T | zeros_{n-r\times1}^T]; \qquad \because x_{ext,n-r\times1} = zeros_{n-r\times1} \qquad (43)$$

Because $r < \min(m, n)$, x is not the unique LS solution of $Ax = p(\vec{b})$. To find the minimum norm solution $x_{min_{norm}}$, B matrix is constructed as expressed in Eq. (14) and x is expressed in terms of B as per Eq. (15) as $\vec{x} = B * \vec{y}$. The above equation can be rewritten as

$$B * \vec{y} = \begin{bmatrix} \Delta_3 * R_c^{-1} * Q_c^{-1} \vec{b} \\ zeros_{n-r\times1} \end{bmatrix} \qquad (44)$$

$$\vec{y} = B^{-1} * \begin{bmatrix} \Delta_3 * R_c^{-1} * Q_c^{-1} \vec{b} \\ zeros_{n-r\times1} \end{bmatrix} \qquad (45)$$

With Δ_1 as defined in the last section, the minimum norm solution $x_{min_{norm}}$ is written as

$$\vec{x}_{min_{norm}} = B * \Delta_1 * B^{-1} * \begin{bmatrix} \Delta_3 * R_c^{-1} * Q_c^{-1} \vec{b} \\ zeros_{n-r\times1} \end{bmatrix} \qquad (46)$$

Hence for this rank deficient case of A, the pseudo inverse is given by

$$A^+ = B * \Delta_1 * B^{-1} * \begin{bmatrix} \Delta_3 * R_c^{-1} * Q_c^{-1} \\ zeros_{n-r\times m} \end{bmatrix} \qquad (47)$$

$$A^+ = B * \Delta_1 * (B^{-1})_{n\times r} * \Delta_3 * R_c^{-1} * Q_c^{-1} \qquad (48)$$

As per Eq. (21), the pseudo inverse of a matrix $A_{m\times n}$ is given by

$$A^+ = B * \Delta_1 * B^{-1} * \Delta_2 * C^{-1}$$

From Eq. (47), we can see that

$$\Delta_2 * C^{-1} = \begin{bmatrix} \Delta_3 * R_c^{-1} * Q_c^{-1} \\ zeros_{n-r\times m} \end{bmatrix} \qquad (49)$$

$$\implies B^{-1} * \Delta_2 * C^{-1} = (B^{-1})_{n\times r} * \Delta_3 * R_c^{-1} * Q_c^{-1} \qquad (50)$$

which implies that Eqs. (48) and (21) are the same.

6 Further Work

In this paper, we presented a geometrical algorithm for computing the generalized inverse of any rank deficient real matrix. The uniqueness of the generalized inverse is also proved. We also presented a scheme of construction for the same. Other variations for special case matrices are also presented. In future, we would like to analyze the special structures and patterns of matrices occurring in specific fields of communications and deep learning where we would like to explore the computational advantages of the proposed algorithm. We would also like to analyze the trade-offs between the computations and the numerical error of the proposed algorithm with respect to the existing algorithms in future.

References

1. Courrieu P (2008) Fast computation of Moore-Penrose inverse matrices. Submitted on May 24, 2005
2. Elaydi SN (2023) Generalized inverses, Drazin inverses, and their applications
3. Recht B, Fazel M, Parrilo PA (2023) A generalized Newton method for solving the matrix rank minimization problem
4. Strang G (2023) Linear algebra and its applications, 3rd ed. Massachusetts Institute of Technology
5. Israel A-B, Greville TNE (2023) Generalized inverses, theory and applications, Canadian mathematical society

Method of Fundamental Solutions for Doubly Periodic Potential Problems

Hidenori Ogata

Abstract In this paper, we propose a method of fundamental solution for two-dimensional doubly periodic problems, especially potential flow problems with a doubly periodic array of obstacles. In the proposed method, we approximate the solution, which involves doubly periodic functions, by a linear combination of the logarithmic potentials consisting of the theta functions. The method inherits the efficiency of the ordinary method of fundamental solutions and gives an approximate solution that has the same periodicity as the one of the exact solution. Numerical examples show the efficiency of the presented method.

Keywords Method of fundamental solutions · Potential problem · Double periodicity · Theta function · Elliptic function

1 Introduction

The method of fundamental solutions is a numerical solver for potential problems

$$
\begin{cases}
\triangle u = 0 & \text{in } \mathscr{D} \\
u = f & \text{on } \partial\mathscr{D},
\end{cases}
\tag{1}
$$

where $\mathscr{D}$ is a domain in the Euclidean space $\mathbb{R}^n$, $\partial\mathscr{D}$ is its boundary, $\triangle$ is the Laplace operator

$$
\triangle = \frac{\partial^2}{\partial x_1^2} + \frac{\partial^2}{\partial x_2^2} + \cdots + \frac{\partial^2}{\partial x_n^2},
$$

Supported by JSPS KAKENHI Grant Number JP21K03366.

H. Ogata (✉)
Department of Computer and Network Engineering, Graduate School of Informatics and Engineering, The University of Electro-Communications, Chofu 182-8585, Japan
e-mail: hidenori.ogata@uec.ac.jp

© The Author(s), under exclusive license to Springer Nature Singapore Pte Ltd. 2024
D. Giri et al. (eds.), *Proceedings of the Tenth International Conference on Mathematics and Computing*, Lecture Notes in Networks and Systems 964,
https://doi.org/10.1007/978-981-97-2066-8_13

and f is a function given on $\partial \mathscr{D}$. In the method, we approximate the solution of (1) by a linear combination of the fundamental solutions of the Laplace operator. In two-dimensional problems ($n = 2$), the approximate solution is of the form

$$u_N(x) = Q_0 - \frac{1}{2\pi} \sum_{j=1}^{N} Q_j \log \|x - \xi_j\|, \tag{2}$$

where N is a positive integer, ξ_j ($j = 1, 2, \ldots, N$) are points given in $\mathbb{R}^2 \setminus \overline{\mathscr{D}}$, $Q_0, Q_1, \ldots, Q_N$ are real coefficients satisfying [1]

$$\sum_{j=1}^{N} Q_j = 0. \tag{3}$$

The coefficients Q_j are called the charges and the points ξ_j are called the charge points. In words of physics, the approximated solution $u_N(x)$ is the superposition of point charges Q_j positioned at points ξ_j. The approximate solution $u_N(x)$ exactly satisfies the Laplace equation $\triangle u_N = 0$. Regarding the boundary condition, we choose the coefficients Q_j for the given points ξ_j so that $u_N(x)$ satisfies the collocation condition, that is, it satisfies the boundary condition at points given on the boundary x_i ($i = 1, 2, \ldots, N$)

$$u_N(x_i) = f(x_i) \quad (i = 1, 2, \ldots, N), \tag{4}$$

where x_i are called the collocation points. Equations (3) and (4) give a system of linear equations for Q_j. We obtain $u_N(x)$ by solving the linear system. Throughout this paper, we identify the two-dimensional Euclidean plane $\mathbb{R}^2$ with the complex plane $\mathbb{C}$, $\mathscr{D} \subset \mathbb{R}^2$ with a complex domain, and a point $(x, y) \in \mathbb{R}^2$ with the complex number $z = x + iy \in \mathbb{C}$. Then, the approximate solution (2) is given as

$$u_N(z) = Q_0 - \frac{1}{2\pi} \sum_{j=1}^{N} Q_j \log |z - \zeta_j|,$$

where ζ_j are the charge points given in $\mathbb{C} \setminus \overline{\mathscr{D}}$, and Q_j are the charges determined by (3) and the collocation equation

$$u_N(z_i) = f(z_i) \quad (i = 1, 2, \ldots, N)$$

with the collocation points $z_1, \ldots, z_N \in \partial \mathscr{D}$.

We can use the method of fundamental solutions to approximate complex analytic functions. Let $f(z)$ be an analytic function on a complex domain $\mathscr{D}$. Its real part

---

[1] This equation is posed for the invariance under scale transformation $u_N(cx) = u_N(x)$ [9].

Re $f(z)$ is a harmonic function, a solution of the Laplace equation on $\mathcal{D}$, and it is approximated using the method of fundamental solutions by

$$\operatorname{Re} f(z) \simeq Q_0 - \frac{1}{2\pi} \sum_{j=1}^{N} Q_j \log |z - \zeta_j|. \tag{5}$$

Then, the imaginary part Im $f(z)$ is its conjugate harmonic function, and it is approximated by

$$\operatorname{Im} f(z) \simeq (\text{ real constant }) - \frac{1}{2\pi} \sum_{j=1}^{N} Q_j \arg(z - \zeta_j). \tag{6}$$

Combining (5) and (6) gives the approximation of f

$$f(z) \simeq f_N(z) = Q_0 - \frac{1}{2\pi} \sum_{j=1}^{N} Q_j \log(z - \zeta_j). \tag{7}$$

From this viewpoint, Amano proposed methods of numerical conformal mapping based on the method of fundamental solutions [2, 3].

In this paper, we examine two-dimensional potential flow problems in doubly periodic domain as shown in Fig. 1. In general, two-dimensional potential flow is described by a complex velocity potential, which is an analytic function $f(z)$ in the flow domain satisfying certain boundary conditions and gives the velocity field $v(z) = (u, v)$ by $u - iv = f'(z)$ [8]. Then, our problem is to find an analytic function which approximates the complex velocity potential. In our problem, the velocity field is obviously a doubly periodic function, and, then, the complex velocity potential $f(z)$ involves doubly periodic functions. It is difficult to approximate such a potential by the conventional method of fundamental solutions, that is, in the form of (7). To overcome this difficulty, we approximate the velocity potential by the superposition of potentials generated by a doubly periodic array of equal charges. In words of mathematics, the potential of a periodic array of charges is given by the composition of the complex logarithmic function and the theta functions, which are entire functions with a doubly periodic array of zeros and satisfies a pseudo-periodicity. It is expected that our method inherits the efficiency of the conventional method of fundamental solutions and satisfies the same periodicity as that of the potential flow of our problem.

Related to this paper, we proposed methods of fundamental solutions for two-dimensional potential problems with one dimensional periodicity [14] and two or three-dimensional Stokes flow problem with two or three-dimensional periodicity [10, 12, 13]. As an application of studies of periodic flow problems, Liron applied a method of periodic Stokes flows to the study of flow transport by cilia [6].

The contents of the paper are as follows. In Sect. 2, we formulate our problem of doubly periodic potential flow. In Sect. 3, we propose our method of fundamental solutions for the problem formulated in the previous section. In Sect. 4, we mention

Fig. 1 The periodic domain
$\mathscr{D}$

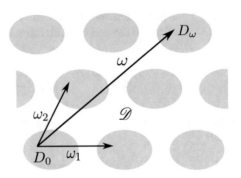

the invariance of our method under the change of the basis of the period lattice. In Sect. 5, we rewrite our formula in terms of Weierstrass' elliptic functions. In Sect. 6, we show a numerical example which shows the effectiveness of our method. In Sect. 7, we give some concluding remarks.

2 Doubly Periodic Potential Flow Problem

First, the flow domain of our problem is given by

$$\mathscr{D} = \mathbb{C} \setminus \bigcup_{\omega \in \Omega} \overline{D}_\omega,$$

where

$$\Omega := \{m\omega_1 + n\omega_2 | m, n \in \mathbb{Z}\}$$

is the period lattice with a basis $\omega_1, \omega_2 \in \mathbb{C} \setminus \{0\}$ such that $\mathrm{Im}(\omega_2/\omega_1) > 0$, and

$$D_\omega = \{z + \omega | z \in D_0\} \quad (\omega \in \Omega)$$

with a bounded simply connected domain D_0 (See Fig. 1). Each D_ω is an obstacle in a periodic array. We examine the potential flow in the domain $\mathscr{D}^2$. The potential flow is described by the complex velocity potential $f(z)$, a complex analytic function in $\mathscr{D}$ whose real part $\Phi = \mathrm{Re}\, f$ and imaginary part $\Psi = \mathrm{Im}\, f$ give the velocity field $\mathbf{v} = (u, v)$ by

$$u = \frac{\partial \Phi}{\partial x} = \frac{\partial \Psi}{\partial y}, \quad v = \frac{\partial \Phi}{\partial y} = -\frac{\partial \Psi}{\partial x}. \tag{8}$$

---

[2] For the detail of two-dimensional potential flow, see [8] for example

Fig. 2 The part of the period parallelogram $\mathscr{D}_0$ outside the obstacles

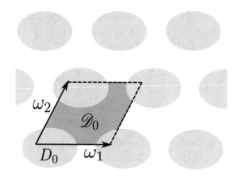

The Eq. (8) are the Cauchy-Riemann relations for Φ and Ψ. From (8), the velocity field is given by

$$u - iv = f'(z).$$

Since we have $0 = d\Psi = \frac{\partial \Psi}{\partial x}dx + \frac{\partial \Psi}{\partial y}dy = -vdx + udy$ along the contour line of Ψ, the contour line $\Psi = \text{constant}$ is a stream line. Therefore, the complex velocity potential that we seek is a complex analytic function $f(z)$ in $\mathscr{D}$ such that it satisfies the boundary condition

$$\text{Im } f = \text{constant} \quad \text{on } \partial\mathscr{D}. \tag{9}$$

In addition, we pose the condition that the mean velocity takes a given value $\langle v \rangle = (U, 0)$

$$\langle f' \rangle = \frac{1}{|\mathscr{D}_0|} \iint_{\mathscr{D}_0} f'(z)dxdy = U, \tag{10}$$

where $\mathscr{D}_0$ is the part of a period parallelogram outside the obstacles, that is,

$$\mathscr{D}_0 = \{a\omega_1 + b\omega_2 | 0 \le a, b \le 1\} \setminus \bigcup_{\omega \in \Omega} \overline{D_\omega},$$

and $|\mathscr{D}_0|$ is its area (See Fig. 2). Therefore, our problem is to find a complex analytic function $f(z)$ in $\mathscr{D}$ which satisfies the boundary condition (9) and the mean velocity condition (10).

3 Method of Fundamental Solutions for Doubly Periodic Potential Flow

The complex velocity potential of our problem $f(z)$ obviously has the periodicity, and, however, it is difficult to approximate such a potential by the conventional method of fundamental solution which gives an approximation using the form (7).

We propose to approximate the potential by

$$f(z) \simeq f_N(z) = Uz - \frac{i}{2\pi} \sum_{j=1}^{N} Q_j \left\{ \log \vartheta_1 \left(\left. \frac{z - \zeta_j}{\omega_1} \right| \tau \right) - v_j z \right\}, \quad (11)$$

where $\zeta_1, \ldots, \zeta_N$ are the charge points given in the obstacle D_0, $Q_1, \ldots, Q_N \in \mathbb{R}$ are the unknown charges satisfying

$$\sum_{j=1}^{N} Q_j = 0, \quad (12)$$

$\vartheta_1(\cdot|\tau)$ is the theta function [3]

$$\vartheta_1(v|\tau) := 2 \sum_{n=0}^{\infty} (-1)^n q^{(n+1/2)^2} \sin(2n+1)\pi v$$

$$= 2q^{1/4} \sin \pi v \prod_{n=1}^{\infty} (1 - q^{2n}) \prod_{n=1}^{\infty} (1 - 2q^{2n} \cos 2\pi v + q^{4n})$$

$$\left(\tau = \frac{\omega_2}{\omega_1}, \ q = e^{i\pi\tau} \right), \quad (13)$$

and $v_1, \ldots, v_N$ are given by

$$v_j = \frac{1}{\omega_1 |\mathcal{D}_0|} \iint_{\mathcal{D}_0} \frac{\vartheta_1'(\pi(z - \zeta_j)/\omega_1|\tau)}{\vartheta_1(\pi(z - \zeta_j)/\omega_1|\tau)} dx dy \quad (j = 1, 2, \ldots, N) \quad (14)$$

so that $f_N(z)$ satisfies the mean velocity condition (10). We determine the unknown charges Q_j by the collocation condition

$$\operatorname{Im} f_N(z_i) = C \text{ (unknown real constant)}$$

for given collocation points $z_1, z_2, \ldots, z_N \in \partial D_0$, that is,

$$-\frac{1}{2\pi} \sum_{j=1}^{N} Q_j \left\{ \log \left| \vartheta_1 \left(\left. \frac{z - \zeta_j}{\omega_1} \right| \tau \right) \right| - \operatorname{Re}(v_j z_i) \right\} - C = -U(\operatorname{Im} z_i)$$

$$(i = 1, 2, \ldots, N). \quad (15)$$

---

[3] For the details of the theta functions, see [4] for example. Remark that $\theta_1(z)$ in [4] is $\vartheta_1(z/\pi|\tau)$ in this paper.

The collocation equation (15) and Eq. (12) form a system of linear equations for the unknown Q_j and C. We obtain the approximate potential $f_N(z)$ by determining the charges Q_j as the solution of the linear system (15) and (12).

The approximate potential (11) is an analytic function in $\mathscr{D}$. In fact, the theta function $\vartheta_1(v|\tau)$ is an entire function with the zeros at $m + n\tau$ ($m, n \in \mathbb{Z}$), and, therefore, $f_N(z)$ is analytic in $\mathbb{C} \setminus \{\zeta_j + \omega | j = 1, 2, \ldots, N, \omega \in \Omega\}$. In addition, the approximate potential $f_N(z)$ satisfies the periodicity which we expect, that is, it satisfies

$$f_N(z + \omega) = f_N(z) + \text{constant} \quad (\forall z \in \mathscr{D}, \forall \omega \in \Omega), \tag{16}$$

and, therefore, the complex velocity $f_N'(z) = u - iv$ satisfies the periodicity

$$f_N'(z + \omega) = f_N'(z) \quad (\forall z \in \mathscr{D}, \forall \omega \in \Omega). \tag{17}$$

In fact, the pseudo-periodicity of the theta function

$$\vartheta_1(v + 1|\tau) = -\vartheta_1(v|\tau), \quad \vartheta_1(v + \tau|\tau) = -q^{-1}e^{-2\pi iv}\vartheta_1(v|\tau)$$

and (12), we obtain $f_N(z + \omega_j) = f_N(z) + \text{constant}$ ($j = 1, 2$).

4   Invariance of the Method

Obviously, the solution of our problem should be independent of the choice of the basis ω_1, ω_2 of the period lattice Ω. However, the approximate solution (11) seems to depend on ω_1 and ω_2 at first sight. Here, we show that the approximation our method gives is invariant under the change of the basis of the period lattice Ω.

Let (ω_1', ω_2') also be a basis of the period lattice Ω. Then, we have the relation

$$\begin{pmatrix} \omega_2' \\ \omega_1' \end{pmatrix} = \begin{pmatrix} a & b \\ c & d \end{pmatrix} \begin{pmatrix} \omega_2 \\ \omega_1 \end{pmatrix}, \tag{18}$$

where $\begin{pmatrix} a & b \\ c & d \end{pmatrix}$ is a matrix belonging to the unimodular group

$$SL(2, \mathbb{Z}) := \{ \begin{pmatrix} a & b \\ c & d \end{pmatrix} | a, b, c, d \in \mathbb{Z}, \, ad - bc = 1 \}. \tag{19}$$

Thus, we have to show the transformed potential

$$\widetilde{f_N}(z) = Uz - \frac{i}{2\pi} \sum_{j=1}^{N} Q_j \left\{ \log \vartheta_1 \left(\frac{z - \zeta_j}{\omega_1'} \middle| \tau' \right) - v_j' z \right\},$$

where

$$\tau' := \frac{\omega_2'}{\omega_1'} = \frac{a\tau + b}{c\tau + d}$$

and

$$v_j' := \frac{1}{\omega_1'|\mathscr{D}_0'|} \iint_{\mathscr{D}_0'} \frac{\vartheta_1'((z - \zeta_j)/\omega_1'|\tau')}{\vartheta_1((z - \zeta_j)/\omega_1'|\tau')} dxdy \quad (j = 1, 2, \ldots, N),$$

$$\mathscr{D}_0' := \{a\omega_1' + b\omega_2'|0 \le a, b \le 1\} \setminus \bigcup_{\omega \in \Omega} \overline{D_\omega},$$

gives the same potential flow as that $f_N(z)$ gives.

It is known that the unimodular group $SL(2, \mathbb{Z})$ is generated by the matrices

$$T = \begin{pmatrix} 1 & 1 \\ 0 & 1 \end{pmatrix}, \quad S = \begin{pmatrix} 0 & -1 \\ 1 & 0 \end{pmatrix},$$

that is, every element of $SL(2, \mathbb{Z})$ is written as a product of expressions of the form T^n ($n \in \mathbb{Z}$) and S^m ($m = 0, 1, 2, 3$) (See Theorem 6.6 of [4], for example). Then, to show the invariance which we need, it suffices to prove that the approximation of our method is invariant under the transformation of the period basis (18) for

$$\begin{pmatrix} a & b \\ c & d \end{pmatrix} = T \text{ or } S.$$

For the matrix T, we have $\omega_1' = \omega_1$, $\omega_2' = \omega_1 + \omega_2$, $\tau' = \tau + 1$. Using $\vartheta_1(v|\tau + 1) = e^{i\pi/4}\vartheta_1(v|\tau)$, which is easily shown from the definition of the theta function, we have $v_j' = v_j$ ($j = 1, 2, \ldots, N$) and $\widetilde{f_N}(z) = f_N(z)$. Therefore, the potential is invariant under the transformation of the matrix T. For the matrix S, we have $\omega_1' = \omega_2$, $\omega_2' = -\omega_1$, $\tau' = -1/\tau$. Using

$$\vartheta_1\left(v\left|-\frac{1}{\tau}\right.\right) = -i(-i\tau)^{1/2} \exp(i\pi\tau v)\vartheta_1(\tau v|\tau),$$

which is known as the Jacobi imaginary transformation (13) (see §4.7 in [4]), and (12), we have

$$v_j' = \frac{1}{\omega_2|\mathscr{D}_0'|} \iint_{\mathscr{D}_0'} \frac{\vartheta_1'((z - \zeta_j)/\omega_2| - 1/\tau)}{\vartheta_1((z - \zeta_j)/\omega_2| - 1/\tau)} dxdy$$

$$= \frac{1}{\omega_2|\mathscr{D}_0'|} \iint_{\mathscr{D}_0} \left\{ \frac{\omega_2}{\omega_1} \frac{\vartheta_1'((z - \zeta_j)/\omega_1|\tau)}{\vartheta_1((z - \zeta_j)/\omega_1|\tau)} + 2\pi i \frac{z - \zeta_j}{\omega_1} \right\} dxdy$$

$$= v_j - \frac{2\pi i \zeta_j}{\omega_1 \omega_2} + c_0,$$

$$c_0 = \frac{2\pi i}{\omega_1 \omega_2 |\mathcal{D}_0|} \iint_{\mathcal{D}_0} z \, dx dy \quad \text{(constant)}$$

and

$$\tilde{f}_N(z) = Uz - \frac{i}{2\pi} \sum_{j=1}^{N} Q_j \left\{ \log \vartheta_1 \left(\frac{z - \zeta_j}{\omega_2} \Big| -\frac{1}{\tau} \right) - v_j' z \right\}$$

$$= Uz - \frac{i}{2\pi} \sum_{j=1}^{N} Q_j \left\{ \log \vartheta_1 \left(\frac{\omega_2}{\omega_1} \frac{z - \zeta_j}{\omega_2} \Big| \tau \right) + i\pi \frac{\omega_2}{\omega_1} \left(\frac{z - \zeta_j}{\omega_2} \right)^2 \right.$$

$$\left. + \log[-i(-i\tau)^{1/2}] - \left(v_j - \frac{2\pi i \zeta_j}{\omega_1 \omega_2} + c_0 \right) z \right\}$$

$$= Uz - \frac{i}{2\pi} \sum_{j=1}^{N} Q_j \left\{ \log \vartheta_1 \left(\frac{z - \zeta_j}{\omega_1} \Big| \tau \right) - v_j z \right\} - \frac{1}{2\omega_1 \omega_2} \sum_{j=1}^{N} Q_j \zeta_j^2$$

$$= f_N(z) + \text{constant},$$

where we used Eq. (12) on the third equality. Thus, the approximate potential is invariant under the transformation of S up to the addition of a constant, and the transformed potential $\tilde{f}_N(z)$ gives the same potential flow as that $f_N(z)$ gives.

Therefore, it is shown that the approximate solution of our method is independent of the choice of the basis of the period lattice Ω.

5   Formulation in Terms of Elliptic Functions

Due to the double periodicity of our problem, it is naturally expected that the solution of our problem is expressed in terms of the elliptic functions, that is, doubly periodic meromorphic functions. In this section, we rewrite our formula in terms of Weierstrass' elliptic functions as in the author's paper [11].

Let $\sigma(z)$ be the Weierstrass sigma function [4]

$$\sigma(z) := z \prod_{\omega \in \Omega \setminus \{0\}} \left(1 - \frac{z}{\omega} \right) \exp \left[\frac{z}{\omega} + \frac{1}{2} \left(\frac{z}{\omega} \right)^2 \right]. \tag{20}$$

It is an entire function with the zeros at periods $\omega \in \Omega$ and satisfies the pseudo-periodicity (Proposition 7.4 in [4])

$$\sigma(u + \omega_j) = -\exp \left(\eta_j \left(u + \frac{\omega_j}{2} \right) \right) \sigma(u) \quad (j = 1, 2)$$

---

[4] For the detail of the elliptic functions, see [1, 4] for example.

with constants η_j given by

$$\eta_j = 2\zeta\left(\frac{\omega_j}{2}\right) \quad (j = 1, 2),$$

where $\zeta(u)$ is the Weierstrass zeta function

$$\zeta(u) := \frac{d}{du}\log\sigma(u) = \frac{1}{u} + \sum_{\omega\in\Omega\setminus\{0\}}\left(\frac{1}{u-\omega} + \frac{1}{\omega} + \frac{u}{\omega^2}\right). \tag{21}$$

The approximate potential $f_N(z)$ given by (11) is rewritten in terms of $\sigma(u)$ as follows. First, we have the formula

$$\vartheta_1\left(\frac{z}{\omega_1}\right) = \frac{\omega_1}{\vartheta_1'}\exp\left(-\frac{\eta_1 z^2}{2\omega_1}\right)\sigma(z) \quad (\vartheta_1' := \vartheta_1'(0|\tau)) \tag{22}$$

(See §19 in [1][5]). Substituting (22) into (11), we have

$$f_N(z) = Uz - \frac{i}{2\pi}\sum_{j=1}^N Q_j\left\{\log\sigma(z-\zeta_j) - \frac{\eta_1(z-\zeta_j)^2}{2\omega_1} - v_j z\right\},$$

where we used Eq. (12). Here, we have to give v_j using the Weierstrass elliptic functions. From (22), we have

$$\frac{1}{\omega_1}\frac{\vartheta_1'(z/\omega_1|\tau)}{\vartheta_1(z/\omega_1|\tau)} = \frac{d}{dz}\log\vartheta_1\left(\frac{z}{\omega_1}\bigg|\tau\right) = \zeta(z) - \frac{\eta_1}{\omega_1}z,$$

and the constants v_j given by (14) is rewritten as

$$v_j = \widetilde{v}_j + \frac{\eta_1\zeta_j}{\omega_1} + c_1,$$

where $\widetilde{v}_1$ and c_1 are constants respectively given by

$$\widetilde{v}_j := \frac{1}{|\mathcal{D}_0|}\iint_{\mathcal{D}_0}\zeta(z-\zeta_j)\mathrm{d}x\mathrm{d}y, \quad c_1 := -\frac{\eta_1}{\omega_1}\frac{1}{|\mathcal{D}_0|}\iint_{\mathcal{D}_0}z\mathrm{d}x\mathrm{d}y. \tag{23}$$

Consequently, using also (12), we have the expression of the approximate potential $f_N(z)$ in terms of Weierstrass' elliptic functions

$$f_N(z) = Uz - \frac{i}{2\pi}\sum_{j=1}^N Q_j\left\{\log\sigma(z-\zeta_j) - \widetilde{v}_j z\right\},$$

---

[5] Remark that, in [1], the basis of the period lattice Ω is written as $(2\omega, 2\omega')$ and ζ_1 is written as 2ζ.

where $\widetilde{v}_j$ are given by (23) and we ignore the addition of a constant. This expression shows clearly the invariance of our approximation under the change of the period basis (ω_1, ω_2) since the definitions of the sigma function (20) and the zeta function (21) are independent of the choice of the period basis.

6   Numerical Example

We show a numerical example which shows the effectiveness of our method. All the computations are performed using programs coded in C++ with double precision.

The example is the problem of flow past a periodic array of cylinders

$$D_0 = \{z \in \mathbb{C} | |z| < r\} \quad (r > 0).$$

We computed the potential flow for $(\omega_1, \omega_2) = (4r, 4ri)$ and $(\omega_1, \omega_2) = (4re^{i\pi/6}, 4ri)$ by our method with the poles $\zeta_j = qr \exp(2\pi i(j-1)/N)$ $(j = 1, 2, \ldots, N;$ $0 < q < 1)$, the collocation points $z_i = r \exp(2\pi i(i-1)/N)$ $(i = 1, 2, \ldots, N)$ and numerical integration of the right-hand side of (14) by the Monte-Carlo method with 1,000,000 random numbers generated by the Mersenne Twister [7]. Figure 3 shows the streamlines of the flows. To estimate the accuracy of our method, we computed the error of the boundary condition (9)

$$\epsilon_N := \max_{i=1,\ldots,256} |\operatorname{Im} f_N(\widetilde{z}_i) - C| \simeq \max_{z \in \partial D_0} |\operatorname{Im} f_N(z) - \operatorname{Im} f(z)|$$

with the equi-distant points $\widetilde{z}_i$ $(i = 1, 2, \ldots, 256)$ on ∂D_0. Figure 4 shows the value ϵ_N as a function of N, where q is the parameter in the assignment of ζ_j. From the

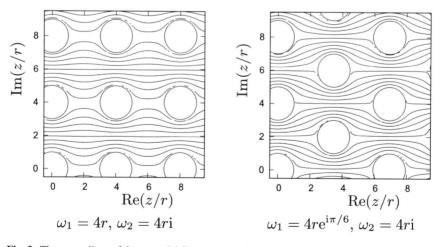

$$\omega_1 = 4r, \ \omega_2 = 4ri \qquad\qquad \omega_1 = 4re^{i\pi/6}, \ \omega_2 = 4ri$$

Fig. 3 The streamlines of the potential flow past a periodic array of cylinders

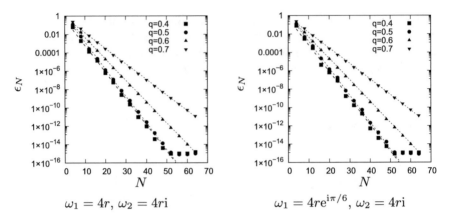

$$\omega_1 = 4r, \quad \omega_2 = 4ri \qquad\qquad \omega_1 = 4re^{i\pi/6}, \quad \omega_2 = 4ri$$

Fig. 4 The error of our method applied to the flow problem past a periodic array of cylinders

Table 1 The decay rates of ϵ_N in the first examples with (1) $(\omega_1, \omega_2) = (4r, 4ri)$ and (2) $(\omega_1, \omega_2) = (4re^{i\pi/6}, 4ri)$

q		0.4	0.5	0.6	0.7
ϵ_N	(1)	$O(0.51^N)$	$O(0.51^N)$	$O(0.58^N)$	$O(0.68^N)$
	(2)	$O(0.51^N)$	$O(0.51^N)$	$O(0.59^N)$	$O(0.68^N)$

figures, we find that ϵ_N decays exponentially as N increases. Table 1 shows the decay rates of ϵ_N estimated by the least square fitting using the command `fit` of the software gnuplot. From the table, we find that ϵ_N approximately obeys

$$\epsilon_N = O(\widetilde{q}^N), \quad \widetilde{q} := \max\{q, 0.5\}.$$

7 Concluding Remarks

In this paper, we proposed a method of fundamental solutions for two-dimensional doubly periodic potential problem, especially, doubly periodic potential flow problems. The flow of our problem is described by a complex analytic function in the flow region called the complex velocity potential which satisfies a pseudo doubly periodicity. In order to approximate the solution which satisfies the double periodicity, we express the approximate solution by a superposition of the complex logarithmic potentials with sources in a doubly periodic array, which is expressed by the theta functions. Our solution is invariant under the change of the basis of the period lattice, and it is rewritten into one expressed in terms of Weierstrass' elliptic functions. A numerical example shows the effectiveness of our method.

A problem related to future studies is an extension of our method for doubly periodic Stokes flow problems. For this problem, the author already proposed a method of fundamental solutions in [13], where we approximate the solution by a superposition of the periodic Stokelets, that is, the Stokes flow with sources in a periodic array. The periodic Stokeslets which the author used in [13] are expressed by a Fourier series. On the other hand, Hasimoto presented a periodic Stokeslet in terms of the elliptic functions [5]. We expect that we can make a method of fundamental solutions for periodic Stokes flow problems using the periodic Stokeslets expressed by the elliptic functions.

References

1. Akhiezer NI (1990) Elements of the theory of elliptic functions. American Mathematical Society, Providence
2. Amano K (1994) A charge simulation method for the numerical conformal mapping of interior, exterior and doubly-connected domains. J Comput Appl Math 53:353–370
3. Amano K (1998) A charge simulation method for numerical conformal mapping onto circular and radial slit domains. SIAM J Sci Comput 19:1169–1187
4. Armitage JV, Eberlein WF (2008) Elliptic functions. Cambridge University Press, Cambridge
5. Hasimoto H (2009) Periodic fundamental solution of the two-dimensional stokes equations. J Phys Soc Japan 78:074401
6. Liron N (1978) Fluid transport by cilia between parallel plates. J Fluid Mech 86:705–716
7. Matsumoto M, Nishimura T (1998) Mersenne twister: a 623-dimensionally equidistributed uniform pseudo-random number generator. ACM Trans Model Comput Simul 8:3–30
8. Milne-Thomson LM (2011) Theoretical hydrodynamics. Dover, New York
9. Murota K (1993) On "invariance" in the schemes in the fundamental solution methods. Trans IPS Japan 34:533–535 in Japanese
10. Ogata H (2006) A fundamental solution method for three-dimensional Stokes flow problems with obstacles in a planar periodic array. J Comput Appl Math 2006:622–634
11. Ogata H (2021) Methods of fundamental solutions for doubly-periodic potential problems. Trans Japan Soc Indust Appl Math 31:1–19 in Japanese
12. Ogata H, Amano K (2006) A fundamental solution method for three-dimensional viscous flow problems with obstacles in a periodic array. J Comput Appl Math 193:302–318
13. Ogata H, Amano K, Sugihara M, Okano D (2003) A fundamental solution method for viscous flow problems with obstacles in a periodic array. J Comput Appl Math 152:411–425
14. Ogata H, Okano D, Amano K (2002) Numerical conformal mapping of periodic structure domains. Japan J Indust Appl Math 19:257–275

Analysis of Multilanguage Regional Music Tracks Using Representation Learning Techniques in Lower Dimensions

Vishnu S. Pendyala⬤, Samhita Konduri, and Kriti V. Pendyala

Abstract Machine understanding of music requires digital representation of the music using meaningful features and then analyzing the features. The work in this paper is unique in using representation learning techniques in lower dimensions for analyzing the effectiveness of mel-spectrogram features of assorted music tracks in multiple languages. The features are plotted in three different transformed feature spaces for visual inspection of the fine-grained attributes of the music rendition such as the vocal artist, their gender, language, and standing in the industry. The analysis of the music tracks in a chosen dataset using spectral and non-spectral algorithms such as Principal Component Analysis (PCA), t-distributed Stochastic Neighbor Embedding (t-SNE), and Uniform Manifold Approximation and Projection (UMAP) provide valuable insights into the representation learning of the selected music tracks. UMAP performs better than the other two algorithms and is able to reasonably discern the various subtler aspects of a music rendition.

Keywords Feature transformation · Mel-spectrogram · Machine Learning · Dimensionality Reduction · Uniform Manifold Approximation and Projection · t-distributed Stochastic Neighbor Embedding

V. S. Pendyala (✉)
Department of Applied Data Science, San Jose State University, 1 Washington Square, San Jose, CA 95192, USA
e-mail: vishnu.pendyala@sjsu.edu

S. Konduri
Palo Alto High School, 50 Embarcadero Rd, Palo Alto, CA, USA
e-mail: samhita.konduri@gmail.com

K. V. Pendyala
University Preparatory Academy, 2315 Canoas Garden Ave, San Jose, CA, USA
e-mail: kriti.v.pendyala@gmail.com

© The Author(s), under exclusive license to Springer Nature Singapore Pte Ltd. 2024
D. Giri et al. (eds.), *Proceedings of the Tenth International Conference on Mathematics and Computing*, Lecture Notes in Networks and Systems 964,
https://doi.org/10.1007/978-981-97-2066-8_14

1 Introduction

Machine learning is increasingly becoming ubiquitous and has been successfully applied to music, specifically for the possibility of building an Indian classical music tutor [18] and for societal progress in general [15]. Machine learning models work on the data representing the information to be analyzed, in this case, music renditions. Representation learning plays a key role in generating value from data. If the representation of data is incorrect or inaccurate, even the most efficient machine learning or other data mining techniques will not be able to generate the expected value from the data. This project therefore sets to answer the following research questions.

- RQ1: What can be concluded about the distributions of the mel-spectrogram features extracted from 30-second snippets of music renditions?
- RQ2: How effective are the representation learning in machine learning used for dimensionality reduction at distinguishing the fine-grained attributes of a rendition such as the artist, their veteran status, language of rendition, and their gender from the mel-spectrogram features so extracted?

Representation learning for this work is a two-stage process. The music tracks are first represented as mel-spectrogram vectors. This representation is high-dimensional because the vectors have many elements. To be able to visually analyze the various aspects of the music tracks, there is a need to generate corresponding embeddings in a lower dimensional space. This work analyzes how well the representations in higher dimensions and the corresponding learned representations in lower dimensions help in discerning fine-grained characteristics of the music tracks.

1.1 Contribution

As the literature survey in Sect. 2 shows, visualization using t-SNE and PCA seems to have been performed in the context of classification and clustering of music but not for an analysis with respect to subtler aspects such as the artist's gender or standing in the industry. As confirmed in this paper, UMAP performs much better than PCA and t-SNE and provides more insights. To the best of our knowledge, this work is unique in performing a detailed spectral and non-spectral analysis of music tracks in various languages to generate key insights regarding the gender, veteran status, the vocalists themselves, and the language of the music track, using an audio representation technique commonly used, the Mel-Spectrogram vectors for music.

2 Related Work

The dimensionality reduction techniques PCA and t-SNE have been used for visualizations of music as a precursor to performing classification [4], discovery and recommendation [11], clustering [12] in multiple papers but not specifically to analyze the strength of representation learning and in comparison with UMAP as is done in this paper. For brevity, this section only lists innovative uses of the algorithms in the context of music. Attempts seem to have been made to use a Robust PCA for separating voice from the ambient music in an audio signal in an unsupervised fashion [3]. Since original PCA [13] is sensitive to outliers and noise, the authors use a robust PCA to separate voice features from a music signal. But as we shall see in this paper, PCA does not perform well at discerning subtle features. T-SNE has been compared with PCA in generating a music playlist based on 34 extracted features such as genres and emotions [6] concluding that t-SNE performs better than PCA. T-SNE and PCA have also been used to visualize embeddings of faces which reflect education in music or its lack thereof [20].

3 Methodology

Traditionally, the first step after collecting raw data is to pre-process and extract features from the data. As described in Sect. 3.1, the dataset procured for the experiments already has the features extracted. Analysis of the data is performed by visualizing the features in lower dimensions. The accuracy of the analysis depends on two primary facets: (a) the representation of the data captured in the features indicating the strength of the representation learning and (b) the effectiveness of the visualization technique in depicting the characteristics of the data. These two facets correspond to the two research questions this paper intends to answer. The following subsections detail the methodology further.

3.1 Dataset

The dataset [21] for the experiments described in this paper is drawn from a larger compilation detailed elsewhere [22]. For the experiments, music tracks in four different languages, Nepali, Nagamese, Kannada, and Marathi, representing the four different regions of the Indian subcontinent, North, East, South, and West respectively are chosen. The dataset for each language is a compressed set of pickle files. Each pickle file contains mel-spectrograms for five songs by that particular vocalist. There are four pickle files in each zip file, one for each of the four different vocalists. Two of these vocalists in each language are male and the other two are female. Two of them are veterans, meaning to have a long standing in the music industry, and two of them are contemporaries who are relatively new to the industry. Some of the songs happen to be duets, so were eliminated from the dataset. Eventually, there are mel-spectrograms for 61 songs in the four pickle files, one for each regional language.

The mel-spectrograms are for three-second segments of the songs that are of a longer duration. Since processing the entire songs is computationally intensive, a random sample of mel-spectrograms for ten consecutive three-second segments is drawn for each song. Each sample is therefore mel-spectrograms for a thirty-second snippet of the music track after concatenating the mel-spectrograms for ten three-second segments. Each thirty-second sample contains 166400 mel-spectrogram features. Visualizing this high-dimensional data with each dimension being a mel-spectrogram value is a challenge that is addressed by performing a representation learning in lower dimensions using spectral and non-spectral analysis.

3.2 Spectral Analysis

High-dimensional data has traditionally been visualized by dimensionality reduction techniques used in machine learning. As stated in Sect. 3.1, the dataset to be visualized for analysis comprises 166400 dimensions, which need to be reduced to two or three dimensions for analyzing using the human eye. Visualization in more than three dimensions cannot be easily comprehended by the human eye. Reducing the dimensions from 166400 to 2 will obviously result in a loss of information. If the information so lost is not significant enough, the visualization is effective.

Often, there is a correlation among the features of the data, in this case, the mel-spectrogram values, which also serve as the dimensions. For instance, consequent samples in an Indian Classical Music (ICM) snippet are likely to be related to each other to create the melody confirming a melodic framework also called as "Raga." Deep learning has been used to convert between melodic frameworks [23]. For processing data, however, correlation implies redundancy. If the information lost by reducing dimensions pertains to this redundancy, then the essence of the data is mostly preserved. When similarities between the various data instances are deemphasized, what matters is the variance among the data.

There are a number of algorithms for generating the embeddings in lower dimensions, a process known as dimensionality reduction. The key idea is to preserve the characteristics of the dataset, even in lower dimensions. The word for "characteristic" in German is "eigen". A dataset is a matrix with rows representing instances, in this case music tracks and columns representing their features, the mel-spectrogram values. The characteristics of a matrix are captured in its eigenvectors. In machine learning, the dimensionality reduction techniques primarily based on computing the eigenvectors are called "spectral methods." Such algorithms have been used for visualizing various aspects of data such as misinformation [17] and fairness [16]. One of the pioneering algorithms in this respect is the PCA, for Principal Component Analysis [13]. PCA captures the essence of the variance in the data by transforming the feature space where the dimensions are completely independent of each other and can be ordered by the magnitude of the variance component in the data each captures.

The primary idea of PCA is to factorize the covariance matrix of the dataset using eigen decomposition. The new features are expressed as a linear combination of the original features where the coefficients are the elements of the eigenvectors. The new features can be ordered by the variance each one captures. For visualization, the top (principal) two features (components) are chosen for the X and Y axis. The terms features, components, and dimensions are used interchangeably and all of them refer to the columns of the original or transformed dataset. For the experiments, the dataset of mel-spectrogram features is first processed using the PCA algorithm to reduce the dimensions to two and plot these top two pseudo-features that capture the most variance in the data.

3.3   Approaches for Non-linear Data

PCA is a linear algorithm. The distances between the data items, expressed as vectors in a vector space, are Euclidean. The assumption therefore is that any two data items are on a hyperplane. This is a simplistic assumption and results in inaccuracies in the representation of the data in the transformed feature space. Particularly knowing that the dataset comprises mel-spectrogram features that are not likely to have a linear relationship among each other, it can be expected that the PCA algorithm may not work well on the music tracks dataset. Another issue with PCA is that it only considers global distances in the form of a covariance matrix and is unconcerned about the local distances among the data items within their natural clusters. This approach does not scale well because of the "curse of dimensionality" [1]. Distances in very high dimensions provide less meaningful sense because the distances are longer and the variance among the distances tends to 0 as the number of dimensions tends to infinity [5]. When the variance is 0, there is no information that the data conveys.

A better approach is to consider the non-linearity in data and preserve only the local distances among the data items such as among the neighbors in a group and forming a cluster. T-SNE, for t-Distributed Stochastic Neighbor Embedding [7] is one such algorithm. T-SNE computes the similarities between each pair of data items based on distances and converts them into probabilities. The goal of t-SNE is to as much as possible, preserve the distribution of the pairwise local distances among neighbors in the original feature space even when the dimensions are reduced, in the new low-dimensional space. This is done by optimizing a cost function that is based on the divergence between the pairwise distances in the high- and low-dimensional spaces, using gradient descent. The mel-spectrograms dataset is next processed using the t-SNE algorithm for better visualization.

With PCA, if the two principal dimensions capture, say 70% of the variance in the 166400 original features of our dataset, 30% of the information in terms of variance is lost. With t-SNE, a significant amount of information pertaining to the global structure is lost. Uniform Manifold Approximation and Projection (UMAP) [8] performs better in these respects. It uses manifold learning to better preserve the global structure of the data and at the same time prioritizing the local distances within groups of data items. UMAP therefore strikes a good balance between retaining global and local topological structures of the data. The next task is therefore to visualize the high-dimensional mel-spectrogram data using UMAP to understand its ability to see through the patterns with respect to the subtler characteristics of the music tracks.

There are several other algorithms for representation learning in lower dimensions. For brevity, only PCA, t-SNE, and UMAP are considered for the experiments owing to their wide acceptance in the scientific community.

3.4 Density Estimation

To answer RQ1 to some extent, it is necessary to visualize the density of the distribution of the mel-spectrograms for the music tracks. Kernel density estimation (KDE) [19] is an effective technique for the purpose. The mel-spectrogram features are first projected to a reduced vector space of two dimensions using one of the foregoing algorithms for dimensionality reduction and then the density is estimated in those two dimensions.

4   Experiments

All of the code for the feature extraction and spectral analysis was written in Python and executed on Google Colab Jupyter Notebooks. The huge dataset is stored in Google Drive and accessed by mounting the drive on Colab. The dataset is well-annotated to indicate the vocalist, whose name is embedded in the pickle file, the number of vocalists in that particular sample, and the name of the music track. As indicated in Sect. 3.1, the mel-spectrogram vectors for ten consequent samples are randomly chosen for each music track and concatenated to make it a 30-second snippet. The vectors so constructed are appended to a data frame in the Pandas [10] package in Python. As a preprocessing step, the dataset is normalized to confirm a unit variance and zero mean distribution using the StandardScaler() method from the scikit-learn package in Python [14].

For dimensionality reduction using PCA, the corresponding class is imported from the same scikit-learn package in Python [14] with the number of dimensions represented by n_components set to 2 and fitted on the normalized dataset. The resulting low-dimensional representation is then plotted using the scatterplot() method of the seaborn package [24]. Similarly, for t-SNE the corresponding method from the scikit-learn package is called with parameters, n_components set to 2, random_state instantiated to 0, and perplexity set to the truncated square root of the number of music tracks in the dataset. Perplexity is a hyperparameter that influences the cluster formation in the low-dimensional space. Its value was chosen heuristically. The t-SNE model is then fitted to the dataset of mel-spectrograms and the resulting representation in two dimensions is plotted using scatterplot().

The t-SNE representations in the lower dimensions can be initialized randomly or using PCA. The representations are improved over several iterations of gradient descent [2]. Figure 1 shows the plots for PCA and t-SNE using various options. As can be seen, t-SNE performs better at preserving local similarities. Similarity is the inverse of the distance between the points. For instance, in Fig. 1d, all three music tracks of the veteran Nepali female vocalists indicated by big green crosses are together, indicating they are similar. It can also be observed that the music track of a contemporary Nepali female vocalist (big green circle) is similar to the tracks of contemporary Marathi female vocalists (big red circles). Similarly, it can be concluded from the plots that two of the tracks by veteran Nagamese female vocalists (big brown cross) are similar to two of the tracks by contemporary Nagamese male singers (small brown circles). The PCA plot on the other hand shows one distinct anomaly, a music track by a contemporary Nepali female vocalist (big green circle to the top right). There is a big green circle appearing a bit different from other points in the t-SNE plots as well, but not as distinct as in the PCA plot.

The UMAP method is imported from umap-learn [9]. the dataset is passed to the model and the generated representation is plotted using scatterplot(). UMAP can work in supervised and unsupervised modes. By default it is unsupervised but can be supervised by passing an optional 'y' value to the UMAP method. Figure 2 illustrates that UMAP is quite successful in discerning the various characteristics of the music

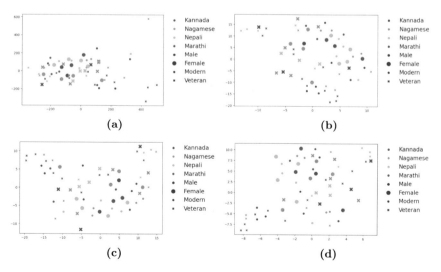

Fig. 1 a PCA plot followed by t-SNE plots initialized by PCA reduced to **b** 5 components **c** 10 components and **d** randomly Each point in the plot represents a music track. Size of the point indicates the gender, color, the language, and shape, the veteran status

tracks. Even within the categories such as the veteran status, the other attributes such as gender and language are meaningfully clustered. For instance, in Fig. 2a, although the music tracks form two very distinct clusters by the vocalist's veteran status, even within those clusters, mini clusters based on gender and language can be seen.

It must be noted that there are multiple stochastic elements in both t-SNE and UMAP algorithms that make them non-deterministic, meaning that it is not guaranteed that repeated runs of the algorithm will result in the same plot. Moreover, the 30-second snippet from the music track is chosen randomly adding to the non-determinism. Also, both the algorithms are highly configurable by means of hyperparameters. The primary hyperparameter in both the algorithm relates to the number of nearest neighbors to be considered in preserving the distances between them. It was observed that the UMAP algorithm performs the best when the n_neighbors parameter is set to 5, which is quite intuitive because the cardinality of most groups of the music tracks with respect to the vocalist, their veteran status, or their gender is around 5. Another observation is that the grouping is best by the vocalist than their gender or veteran status, which is also quite intuitive because songs of most vocalists tend to be similar compared to the music tracks of the artists by gender or veteran status.

Given that UMAP performs better than the other two, to further analyze the finer grained attributes, UMAP plots are generated for each language in a supervised manner based on gender, language, and veteran status and then in an unsupervised mode without providing any labels. As can be seen from Figs. 3, 4, 5, and 6, individual language plots also show discernible characteristics grouped together. For instance, in Fig. 4a, the vocalist, Vaishali Samant's music tracks (big red crosses) are almost

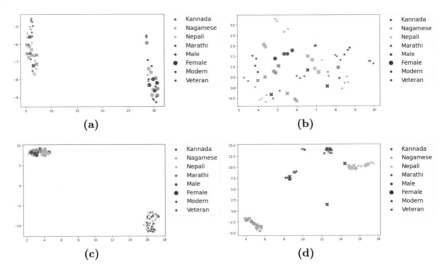

Fig. 2 UMAP plots supervised by **a** veteran status **b** vocalist **c** gender **d** language As in Fig. 1, each point in the plot represents a music track. The size of the point indicates the gender, color, language, and shape, of the veteran status

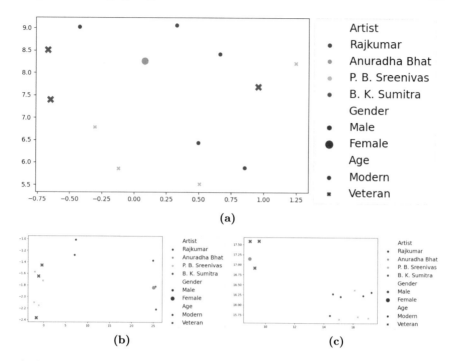

Fig. 3 UMAP plots for Kannada music tracks **a** unsupervised and supervised by **b** veteran status **c** vocalist's gender color indicates the vocalist, size indicates their gender and shape and their veteran status

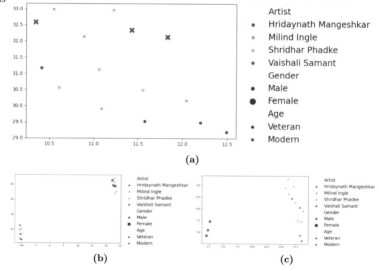

Fig. 4 UMAP plots for Marathi music tracks **a** unsupervised and supervised by **b** veteran status **c** vocalist's gender color indicates the vocalist, size indicates their gender and shape and their veteran status

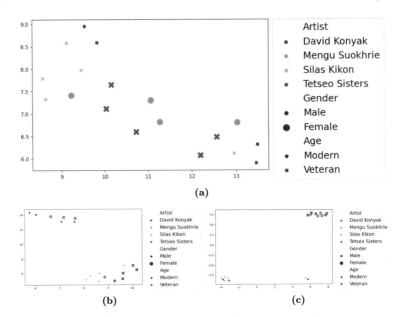

Fig. 5 UMAP plots for Nagamese music tracks **a** unsupervised and supervised by **b** veteran status **c** vocalist's gender color indicates the vocalist, size indicates their gender and shape and their veteran

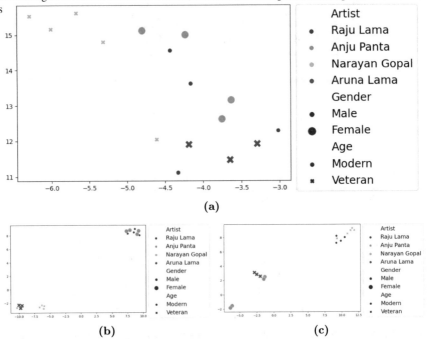

Fig. 6 UMAP plots for Nepali music tracks **a** unsupervised and supervised by **b** veteran status **c** vocalist's gender color indicates the vocalist, size indicates their gender and shape and their veteran status

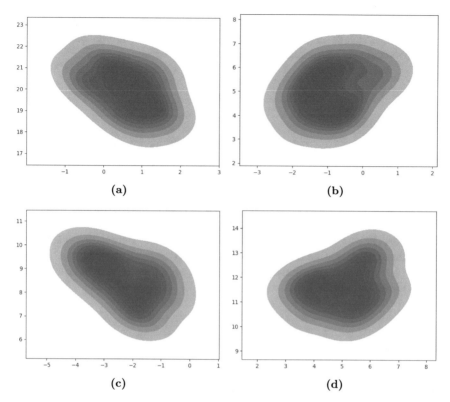

Fig. 7 KDE plots for music tracks in **a** Kannada **b** Marathi **c** Nagamese **d** Nepali

parallel to Milind Ingle's tracks, probably because they are both contemporary (modern) vocalists. In terms of diversity, Hridaynath Mangeshkar's songs cover a broad spectrum in the figure and indeed, a web search shows that the vocalist has rendered diverse forms of music. Similar conclusions can be drawn from the plots for the other languages as well.

To estimate the density of the distribution of the mel-spectrograms, the transformed features in two dimensions generated by the call to UMAP are provided as inputs to the kdeplot() method in the seaborn package. As can be seen from Fig. 7, the densities for each of the languages are quite distinct in their concentrations, shapes, and scales.

Based on the above plots and observations, the research questions can be answered as follows.

- RQ1: The mel-spectrogram features extracted from 30-second snippets of music renditions provide reasonable insights into the subtler aspects of the music track, provided the algorithms running on top of the representation are able to leverage the discernible features in the mel-spectrogram vectors.
- RQ2: The representation learning for dimensionality reduction varies substantially in their performance at distinguishing the fine-grained attributes of a rendition

such as the vocalist, their veteran status, language of rendition, and their gender. UMAP specifically performs better than t-SNE and PCA and the categorization by the vocalist is best compared to any other categorization.

5 Conclusion

The application of AI/ML in the music domain is rapidly progressing. Representation learning plays an important role in any Data Mining or AI/ML application. In this paper, a variety of representation learning techniques in lower dimensions were used to examine the music tracks with respect to finer grained attributes such as the vocalist's gender, veteran status, and language. Among the three algorithms used for the purpose, PCA, t-SNE, and UMAP, discernible attributes were more evident using UMAP and categorization is best by vocalist. To further determine the effectiveness, various classification algorithms will be used on top of the representation learning schemes as a future study.

References

1. Bellman R (1957) Dynamic programming. Princeton University Press
2. Cauchy AL (1847) Méthode générale pour la résolution des systèmes d'équations simultanées. Comptes rendus hebdomadaires des séances de l'Académie des sciences 25:536–538
3. Huang PS, Chen SD, Smaragdis P, Hasegawa-Johnson M (2012) Singing-voice separation from monaural recordings using robust principal component analysis. In: 2012 IEEE international conference on acoustics, speech and signal processing (ICASSP). IEEE, pp 57–60
4. Kumar A, Solanki SS, Chandra M (2022) Stacked auto-encoders based visual features for speech/music classification. Expert Syst Appl 208:118041
5. Lellouche S, Souris M (2019) Distribution of distances between elements in a compact set. Stats 3(1):1–15
6. Lionello M, Pietrogrande L, Purwins H, Abou-Zleikha M (2018) Interactive exploration of musical space with parametric t-sne, pp 200–207 (2018), cited by: 1
7. Van der Maaten L, Hinton G (2008) Visualizing data using t-sne. J Mach Learn Res 9(11)
8. McInnes L, Healy J, Melville J (2018) Umap: uniform manifold approximation and projection. J Open Sour Softw 3(29):861
9. McInnes L, Healy J, Saul N, Grossberger L (2020) Umap: uniform manifold approximation and projection. J Open Sour Softw 5(45):1980
10. McKinney W (2010) Data structures for statistical computing in python. In: Proceedings of the 9th python in science conference, vol 445, pp 51–56
11. Melchiorre AB, Penz D, Ganhör C, Lesota O, Fragoso V, Fritzl F, Parada-Cabaleiro E, Schubert F, Schedl M (2023) Emotion-aware music tower blocks (emomtb): an intelligent audiovisual interface for music discovery and recommendation. Int J Multimedia Inf Retrieval 12(1):13
12. Pavitha N, Khanwelkar D, More H, Soni N, Rajani J, Vaswani C (2022) Analysis of clustering algorithms for music recommendation. In: 2022 IEEE 7th international conference for convergence in technology (I2CT). IEEE, pp 1–6
13. Pearson K (1901) Liii on lines and planes of closest fit to systems of points in space. The London, Edinburgh, and Dublin Philos Magazine J Sci 2(11):559–572

14. Pedregosa F, Varoquaux G, Gramfort A, Michel V, Thirion B, Grisel O, Blondel M, Pretten-hofer P, Weiss R, Dubourg V, Vanderplas J, Passos A, Cournapeau D, Brucher M, Perrot M, Duchesnay É (2011) Scikit-learn: machine learning in python. J Mach Learn Res 12:2825–2830
15. Pendyala VS (2022) Machine learning for societal improvement, modernization, and progress. IGI Global
16. Pendyala VS, Kim H (2023) Analyzing and addressing data-driven fairness issues in machine learning models used for societal problems. In: 2023 international conference on computer, electrical and communication engineering (ICCECE). IEEE, pp 1–7
17. Pendyala VS, Tabatabaii FSA (2023) Spectral analysis perspective of why misinformation containment is still an unsolved problem. In: 2023 IEEE conference on artificial intelligence (CAI). IEEE, pp 210–213
18. Pendyala VS, Yadav N, Kulkarni C, Vadlamudi L (2022) Towards building a deep learning based automated Indian classical music tutor for the masses. Syst Soft Comput 4:200042
19. Rosenblatt M (1956) Remarks on some nonparametric estimates of a density function. Ann Math Stat 832–837
20. Senichenkova YO, Polyak MD (2023) Finding traces of music education on facial images. In: 2023 XXVI international conference on soft computing and measurements (SCM). IEEE, pp 167–170
21. Singh Y, Biswas A (2021) Indian regional music dataset. https://doi.org/10.5281/zenodo.5825830
22. Singh Y, Biswas A (2021) Multitask learning based deep learning model for music artist and language recognition. In: Proceedings of the workshop on speech and music processing 2021, pp 20–23
23. Surana, R, Varshney, A, Pendyala, V (2012) Deep learning for conversions between melodic frameworks of Indian classical music. In: Proceedings of second international conference on advances in computer engineering and communication systems: ICACECS 2021. Springer, pp 1–12
24. Waskom M (2021) Seaborn: statistical data visualization. J Open Sour Softw 6(60):3021

On Some Sequence Spaces in Fuzzy

Ekrem Savas

Abstract In this paper, our main aim is to investigate some new sequence spaces in fuzzy. For this, we shall construct some very interesting new sequence spaces by using the infinite matrix and further some relations between sequence spaces are presented. And as far as we know, our results are new.

Keywords Fuzzy numbers · Almost convergence · Infinite matrix

Classifications 40C45

1 Introduction and Preliminaries

Georg Cantor, a German mathematician, developed set theory, an important branch in Mathematics. A collection of distinct and unambiguous objects is known as a set. When the description on the object is not well defined then uncertainty arises. Due to this point of view, Zadeh [11] introduces fuzzy set in 1965 and in the years that followed, he studied fuzzy logic [12] and fuzzy algorithms [13]. Fuzzy theory plays an important role in the field of mathematics, engineering, logic, computer science, medical science, image processing, and many others. However, no researcher discussed the concept of sequence of fuzzy numbers until Matloka. Matloka [5] considered the sequences of fuzzy numbers and defined the limit of sequence of fuzzy numbers. Further he presents the basic properties of limit. Later the sequences of fuzzy numbers are studied by Nanda [7] and it was shown that the set of all convergent sequences of fuzzy number forms is complete metric space. Subsequently many mathematicians studied sequences of fuzzy numbers.

In this paper, we consider some new convergent fuzzy sequence spaces, using the regular matrix and furthermore we investigate some basic topological properties. For terminology in fuzzy sets and numbers, we refer to [1] and Matloka [5].

E. Savas (✉)
Department of Mathematics, Uşak University, Uşak, Turkey
e-mail: ekremsavas@yahoo.com

Let U be the set of all bounded intervals $\alpha = \left[\bar{\alpha}, \underline{\alpha}\right]$ on the real line $\mathbb{R}$. For $\alpha, \beta \in U$, define

$$\alpha \leq \beta \text{ if and only if } \underline{\alpha} \leq \underline{\beta} \text{ and } \bar{\alpha} \leq \bar{\beta},$$

$$d(\alpha, \beta) = \max\left(\underline{\alpha} - \underline{\beta}, \bar{\alpha} - \bar{\beta}\right).$$

Then it is known that d defines a metric on U (see [1]) and (U, d) is a complete metric space.

A fuzzy number is a fuzzy subset of the real line $\mathbb{R}$ such that bounded, convex, and normal. Let $L(\mathbb{R})$ denote the set of all fuzzy numbers such that upper semi-continuous and have compact support, i.e., if $E \in L(\mathbb{R})$ then for any $\eta \in [0, 1]$, E^η is compact where

$$E^\eta = \begin{pmatrix} t : E(t) \geq \eta \text{ if } 0 < \eta \leq 1, \\ t : E(t) \geq 0 \quad \text{if } \eta = 0. \end{pmatrix}$$

For each $0 < \eta \leq 1$ the $\eta-$level set E^η is a nonempty compact subset of $\mathbb{R}$. The linear structure of $L(\mathbb{R})$ includes addition $E + F$ and scalar multiplication ρE, (ρ a scalar) in terms of $\eta-$level sets by

$$[E + X]^\eta = [E]^\eta + [X]^\eta \text{ and } [\rho E]^\eta = \rho [E]^\eta,$$

each for $0 \leq \eta \leq 1$.

Define a map $\bar{d}(E, X) = \sup_{0 \leq n \leq 1} d(E^\eta, X^\eta)$.

For $E, X \in L(\mathbb{R})$ define $E \leq X$ if and only if $E^\eta \leq X^\eta$ for $\eta \in [0, 1]$. It is obvious that $\left(L(\mathbb{R}), \bar{d}\right)$ is complete metric space (see [7]).

We now review some basic definitions and notation of fuzzy sequences (see [9, 10]).

Definition 1 Suppose that $\xi = (\xi_k)$ is a sequence of fuzzy numbers. If for every $\varepsilon > 0$ there will exist a positive integer N_0 such that

$$\bar{d}(\xi_k, \xi_0) < \epsilon \quad \text{for} \quad k > N_0,$$

we say that the sequence $\xi = (\xi_k)$ is convergent to ξ_0, written as $lim_k \xi_k = \xi_0$.

Let $c(F)$ denote the collection of all convergent sequences of fuzzy numbers.

Definition 2 Suppose that $\xi = (\xi_k)$ is a sequence of fuzzy numbers. If for every $\varepsilon > 0$ there will exist a positive integer N_0 such that

$$\bar{d}(\xi_k, \xi_m) < \epsilon \quad \text{for} \quad k, m > N_0,$$

we say that a sequence $\xi = (\xi_k)$ is a Cauchy sequence. Let $C(F)$ denote the collection of all Cauchy sequences of fuzzy numbers.

Definition 3 The space $\ell_\infty(F)$ is a collection of all fuzzy numbers such that the set $\{\xi_k : k \in N\}$ of fuzzy numbers is bounded.

In [7] it was shown that $c(F)$ and $\ell_\infty(F)$ are complete metric spaces.

By a paranorm we mean a function $\gamma : A \to \mathbb{R}$ (where A is a linear space) which satisfies the following conditions:

p.1 $\gamma(0) = 0$,
p.2 $\gamma(a) \geq 0$ for all $a \in A$,
p.3 $\gamma(-a) = \gamma(a)$ for all $a \in A$,
p.4 $\gamma(a + b) \leq \gamma(a) + \gamma(b)$ for all $a, b \in A$,
p.5 If (μ_n) is sequence of scalar with $\mu_n \to \mu(n \to \infty)$ and (a_n) is a sequence of the elements of A with $\gamma(a_n - a) \to 0(n \to \infty)$, then $g(\mu_n a_n - \mu a) \to 0(n \to \infty)$.

The space α is called the paranormed space with the paranorm γ.

2 Sequence Spaces in Fuzzy

Recently the following space of sequences of fuzzy numbers was defined by Nuray and Savas [8].

$$\ell(q) = \left\{ \xi = (\xi_k) : \sum_k [\overline{d}(\xi_k, 0)]^{q_k} < \infty \right\},$$

where (q_k) is a bounded sequence of strictly positive real numbers. If $q_k = q$ for all k, then $\ell(q) = \ell_q$. Various types of sequence spaces have been presented as follows:

Suppose $T = (t_{nk}(i))$ is the generalized three parametric regular matrix. We write

$$F_0[T, q] = \left\{ \xi = (\xi_k) : \lim_n \sum_k t_{nk}(i)[(\overline{d}(\xi_k, 0)]^{q_k} = 0, \text{ uniformly in } i \right\},$$

$$F[T, p] = \left\{ \xi = (\xi_k) : \lim_n \sum_k t_{nk}(i)[(\overline{d}(\xi_k, \xi_0)]^{q_k} = 0, \text{ uniformly in } i \right\},$$

$$F_\infty[T, p] = \left\{ \xi = (\xi_k) : \sup_{n,i} \left(\sum_k t_{nk}(i)[(\overline{d}(\xi_k, 0)]^{q_k} \right) < \infty \right\},$$

we call them the spaces of strongly T-convergent to 0, strongly T-convergent to ξ_0, and strongly T-bounded sequences of fuzzy numbers $\xi = (\xi_k)$, respectively. We can get some particular cases of these sequence spaces as follows. For all i,

(i) If $t_{nk}(i) = \begin{cases} 1, & 1 \le k \le n, \\ 0, & k > n, \end{cases}$.

Then we obtain the following sequence spaces:

$$F_0(q) = \left\{ \xi = (\xi_k) : \lim_v \frac{1}{v} \sum_{k=1}^{v} [(\overline{d}(\xi_k, 0)]^{q_k} = 0 \right\},$$

$$F(p) = \left\{ \xi = (\xi_k) : \lim_v \frac{1}{v} \sum_{k=1}^{v} [(\overline{d}(\xi_k, \xi_0)]^{q_k} = 0 \right\},$$

$$F_\infty(p) = \left\{ \xi = (\xi_k) : \sup_v \frac{1}{v} \sum_{k=1}^{v} [(\overline{d}(\xi_k, 0)]^{q_k} < \infty \right\},$$

which are called the spaces of strongly convergent to 0, strongly convergent to ξ_0, and strongly bounded, respectively (see [6]).

(ii) If $T = I$, another collection of the sequence spaces for fuzzy numbers is formulated, i.e.,

$$F(q) = c(F, q) = \left\{ \xi = (\xi_k) : [(\overline{d}(\xi_k, \xi_0)]^{q_k} \to 0 (n \to \infty) \right\},$$

$$F_0(q) = \left\{ \xi = (\xi_k) : [(\overline{d}(\xi_k, 0)]^{q_k} \to 0 (n \to \infty) \right\},$$

$$F_\infty(q) = \left\{ \xi = (\xi_k) : \sup_k [(\overline{d}(\xi_k, 0)]^{q_k} < \infty \right\}.$$

(iii) If $q_k = q$ for all k, we have

$$F_0^q = \left\{ \xi = (\xi_k) : \lim_v \frac{1}{v} \sum_{k=1}^{v} t_{nk}(i) [(\overline{d}(\xi_k, 0)]^q = 0, \text{ uniformly in } i \right\},$$

$$F^q = \left\{ \xi = (\xi_k) : \lim_v \frac{1}{v} \sum_{k=1}^{v} t_{nk}(i) [(\overline{d}(\xi_k, \xi_0)]^q = 0, \text{ uniformly in } i \right\},$$

$$F_\infty^q = \left\{ \xi = (\xi_k) : \sup_{v,i} \frac{1}{v} \sum_{k=1}^{v} t_{nk}(i) [(\overline{d}(\xi_k, 0)]^q < \infty \right\}.$$

(iii) If

$$t_{nk}(i) = \frac{1}{i+1} \sum_{j=0}^{i} t_{n+j,k}$$

then we obtain

$$F_0[\hat{T}, q] = \left\{ \xi = (\xi_k) : \lim_i \frac{1}{i+1} \sum_{j=0}^{i} t_{n+j,k} [(\overline{d}(\xi_k, 0)]^{q_k} = 0, \text{ uniformly in } j \right\},$$

$$F[\hat{T}, q] = \left\{ \xi = (\xi_k) : \lim_i \frac{1}{i+1} \sum_{j=0}^{i} t_{n+j,k} [(\overline{d}(\xi_k, \xi_0)]^{q_k} = 0, \text{ uniformly in } j \right\},$$

$$F_\infty[\hat{T}, q] = \left\{ \xi = (\xi_k) : \sup_{i,j} \left(\frac{1}{i+1} \sum_{j=0}^{i} t_{n+j,k} [(\overline{d}(\xi_k, 0)]^{q_k} \right) < \infty \right\},$$

we call them the spaces of strongly almost T-convergent to zero, strongly almost T-convergent to ξ_0, and strongly almost T-bounded sequences, respectively. If we consider $T = (t_{nk}^i)$, for all i,

$$t_{nk}^i := \left\{ \begin{array}{ll} \frac{1}{\sigma_n}, & \text{if } k \in J_n = [n - \sigma_n + 1, n], \\ 0, & \text{otherwise.} \end{array} \right\}$$

such that (σ_n) is a non-decreasing sequence of positive numbers tending to ∞ and $\sigma_{n+1} \leq \sigma_n + 1$, $\sigma_1 = 1$ (see [9]), so these definitions are reduced to following, for all i

$$F[V, \sigma, q] = \left\{ x : \lim_n \frac{1}{\sigma_n} \sum_{k \in J_n} [(\overline{d}(\xi_k, \xi_0)]^{q_k} = 0, \right\}$$

and

$$F[V, \sigma, q]_0 = \left\{ x : \lim_n \frac{1}{\sigma_n} \sum_{k \in J_n} [(\overline{d}(\xi_k, 0)]^{q_k} = 0, \right\}$$

$$F[V, \sigma, q]_\infty = \left\{ \xi = (\xi_k) : \sup_n \frac{1}{\sigma_n} \sum_{k \in J_n} [(\overline{d}(\xi_k, 0)]^{q_k} \right) < \infty \right\}.$$

When $\sigma_n = n$ for all n, the above sequence spaces become $F(q)$, $F_0(q)$ and $F_\infty(q)$.

If $T = (t_{nk}^i)$ is considered as for all i

$$t_{nk}^i := \left\{ \begin{array}{ll} \frac{1}{v_l}, & \text{if } k_{l-1} < k \leq k_l, \\ 0, & \text{otherwise,} \end{array} \right\}$$

where $\vartheta = (k_l)$ is an increasing integer sequence such that $k_0 = 0$ and $v_l = k_l - k_{l-1} \to \infty$, as $l \to \infty$. Let $I_l = (k_{l-1}, l_l]$ (see [2, 3]). then these definitions become, for all i,

$$F[N_\vartheta, q] = \left\{ \xi : \lim_l \frac{1}{v_l} \sum_{k \in I_l} [(\overline{d}(\xi_k, \xi_0)]^{q_k} = 0, \right.$$

and

$$F[N_\vartheta, q]_0 = \left\{ \xi : lim_l \frac{1}{v_l} \sum_{k \in I_l} [(\overline{d}(\xi_k, 0)]^{q_k} = 0, \right.$$

$$F[N_\vartheta, q]_\infty = \left\{ \xi = (\xi_k) : \sup_l \frac{1}{v_l} \sum_{k \in I_l} [(\overline{d}(\xi_k, 0)]^{q_k} \right) < \infty \right\}.$$

Proposition 4 (See, [6]). *If $\overline{d}$ is a translation invariant metric on $L(\mathbb{R})$, so*

(i) $\overline{d}(E + F, 0) \le \overline{d}(E, 0) + \overline{d}(F, 0)$,
(ii) $\overline{d}(\mu E, 0) \le |\mu| \overline{d}(E, 0), |\mu| > 1$.

When $\overline{d}$ is a translation invariant, we obtain the following result.   □

Proposition 5 *Suppose that (q_k) is a bounded sequence of strictly positive real numbers. Then $F_0[T, q]$, $F[T, q]$ and $F_\infty[T, q]$ are linear spaces over the complex field $\mathbb{C}$.*

Proposition 6 *$F_0[T, q]$, $F[T, q]$ and $F_\infty[T, q]$ are absolutely convex subsets of the space $w(F)$ of all sequences of fuzzy numbers, where $0 < q_k \le 1$.*

3   Main Results

Theorem 7 *$F_0[T, q]$ and $F[T, q]$ are complete paranormed spaces equipped with the paranorm g which is*

$$g(E) = \sup_{n,i} \left(\sum_k t_{nk}(i)[\overline{d}(\xi_k, 0)]^{q_k} \right)^{1/M}$$

where $M = \max(1, \sup_k q_k)$.

Proof Obviously $g(\theta) = 0$, $g(-\xi) = g(\xi)$. It is clear that $g(\xi + \rho) \le g(\xi) + g(\rho)$ for $\xi = (\xi_k)$, $\rho = (\rho_k)$ in $F_0[T, q]$. For arbitrary scalar μ, $|\mu|^{q_k} < \max\{1, |\mu|^H\}$, where $H = sup_k q_k < \infty$ is obtained, so

$$g(\mu, \xi) < (\sup_k |\mu|^{q_k})^{1/M} . g(\xi) \text{ on } F_0[T, q].$$

Therefore $\mu \to 0, \xi \to \theta \Rightarrow \mu\xi \to \theta$ and further $\xi \to \theta, \mu$ fixed $\Rightarrow \mu\xi \to \theta$. Suppose $\mu \to 0, \xi$ fixed. We write

$$\sum_k t_{nk}(i)[\overline{d}(\mu\xi_k, 0)]^{q_k} < \epsilon \text{ for } n > N(\epsilon),$$

for $|\mu| < 1$. Further, for $1 \le n \le N$, since $\sum_k t_{nk}(i)[\overline{d}(\xi_k, 0)]^{q_k} < \infty$, there will exist m which are

$$\sum_{k=m}^{\infty} t_{nk}(i)[\overline{d}(\mu\xi_k, 0)]^{q_k} < \epsilon.$$

Taking μ small enough then we get

$$\sum_k t_{nk}(i)[\overline{d}(\mu\xi_k, 0)]^{q_k} < 2\epsilon \text{ for all } n.$$

Finally $g(\mu\xi) \to 0$ when $\mu \to 0$. Hence g is a paranorm on $F_0[T, q]$. We can prove the completeness by following the same lines as in [10] for $\ell(p)$.

The space $F[T, q]$ has exactly the same proof. □

Theorem 8 *If* $0 < |\inf_k q_k \le \sup_k q_k < \infty$, *then* $F_\infty[T, q]$ *is a paranormed space with the above paranorm.*

The proof is obvious.

Theorem 9 *Suppose* $0 < q_k \le r_k$ *and* (r_k/q_k) *is bounded. Then* $F[T, r] \subseteq F[T, q]$.

Proof Assume $\xi = (\xi_k) \in F[T, q]$, write $v_k = [\overline{d}(\xi_k, \xi_0)]^{r_k}$ and $\mu_k = r_k/q_k$. Of course $0 < \mu_k \le 1$. Consider $0 < \mu < \mu_k$. Define, $w_k = \begin{cases} v_k, & v_k \ge 1, \\ 0, & v_k < 1, \end{cases}$, and $z_k = \begin{cases} 0, & v_k \ge 1, \\ v_k, & v_k < 1, \end{cases}$.

Then we get $v_k = w_k + z_k$ and $v_k^{\mu_k} = w_k^{\mu_k} + z_k^{\mu_k}$ and it is obvious that $w_k^{\mu k} \le w_k \le v_k$ and $z_k^{\mu k} \le z_k^{\lambda}$. Thus,

$$\sum_k t_{nk}(i)[\overline{d}(\xi_k, \xi_0)]^{q_k} = \sum_k t_{nk}(i)v_k^{\mu k} = \sum_k t_{nk}(i)(a_k^{\mu k} + z_k^{\mu k}) =$$

$$\sum_k a_{nk}(i)v_k + \sum_k t_{nk}(i)z_k^{\mu} \to 0(n \to \infty),$$

uniformly in i. Since $\xi \in F[T, r]$, $\sum_k t_{nk}(i)v_k$ is convergent, and $z_k < 1$ and T is regular. $\sum_k t_{nk}(i)z_k^{\mu}$ is also convergent. As a result $\xi \in F[T, q]$, i.e., $F[T, r] \subseteq F[T, q]$. □

Theorem 10 *Let n_1 and n_2 be constants which are $0 < n_1 \leq q_k \leq n_2$. Then $\xi \in c(F)$ implies $\xi \in F[T, q]$ with $\lim_k \xi_k = F[T, q] - \lim_k \xi_k = \xi_0$ if and only if $T = (t_{nk}(i))$ $T \in (c_0(F), c_0(F))$.*

Proof Sufficiency: Since $q_k \geq n_1 > 0$, we get

$$[\overline{d}(\xi_k, \xi_0)] \to 0 \implies [\overline{d}(\xi_k, \xi_0)]^{q_k} \to 0$$

thus $T \in (c_0(F), c_0(F))$ implies that $\sum_k t_{nk}(i)[\overline{d}(\xi_k, \xi_)]^{q_k} \to 0(n \to \infty)$, i.e., $\xi \in c(F) \implies \xi \in F[T, q]$ with the same limit ξ_0.

Necessity: Suppose

$$[\overline{d}(\xi_k, \xi_0)] \to 0 \implies \sum_k t_{nk}(i)[\overline{d}(\xi_k, \xi_0)]^{q_k} \to 0(n \to \infty),$$

uniformly in i. Then

$$[\overline{d}(\xi_k, \xi_0)]^{r_k} \to 0(k \to \infty) \implies \sum_k t_{nk}(i)[\overline{d}(\xi_k, \xi_0)] \to 0, (n \to \infty), \tag{1}$$

uniformly in i, where $r_k = 1/q_k$. Since $r_k \geq 1/n_2 > 0$,

$$\overline{d}(\xi_k, \xi_0) \to 0(k \to \infty) \implies [\overline{d}(\xi_k, \xi_0)]^{r_k} \to 0(n \to \infty), \tag{2}$$

uniformly in i.

Therefore by (1) and (2), we get

$$\overline{d}(\xi_k, \xi_0)] \to 0(k \to \infty) \implies \sum_k t_{nk}(i)[\overline{d}(\xi_k, \xi_0)] \to 0, (n \to \infty),$$

uniformly in i.

Hence $T \in (c_0(F), c_0(F))$. $\qquad\qquad\qquad\qquad\qquad\qquad\qquad\qquad\square$

References

1. Diamond P, Kloeden P (1990) Metric spaces of fuzzy sets. Fuzzy Sets Syst 35:241–249
2. Freedman AR, Sember JJ, Raphel M (1978) Some cesaro-type summability spaces. Proc Lond Math Soc 37:508–520
3. Kwon JS, Shim HT (2001) Remark on Lacunary statistical convergence of fuzzy numbers. Fuzzy Sets Syst 123:85–88
4. Maddox IJ (1970) Elements functional analysis. Cambridge Univ, Press
5. Matloka M (1986) Sequences of fuzzy numbers. BUSEFAL 28:28–37
6. Mursaleen, Basarir M (2003) On some new sequence spaces of fuzzy numbers. Indian J Pure Appl Math 34(9):1351–1357
7. Nanda S (1989) On sequence of fuzzy numbers. Fuzzy Sets Syst 33:123–126

8. Nuray F, Savaş E (1995) Statistical convergence of sequences of fuzzy numbers. Math Slovaca 45(3):269–273
9. Savaş E (2000) On strongly λ-summable sequences of fuzzy numbers. Inform Sci 125:181–186
10. Savaş E (2000) A note on sequence of fuzzy numbers. Inform Sci 124:297–300
11. Zadeh A (1965) Fuzzy sets. Infor Control 8:335–338
12. Zadeh A (1966) Fuzzy logic. IEEE Trans Fuzzy Syst 4:103–111
13. Zadeh A (1968) Fuzzy algorithms. Inform Control 12:94–102

Stabilizer Group of Set Ideals

S. B. Ramkumar[ID] **and V. Renukadevi**[ID]

Abstract We derive a necessary and sufficient condition for the isomorphism between stabilizer groups of two set ideals on an infinite set. Also, we prove that the outer automorphism group of stabilizer group of set ideal is a group of order at most two if the ideal is isomorphic to its polar.

Keywords Ideal · Symmetric group · Complete · Automorphism · Cardinal · Lagrange ideal

1 Introduction

Let S denote the symmetric group on X. The *stabilizer group* of an ideal $\mathcal{I}$ on a nonempty set X is the stabilizer of $\mathcal{I}$ in S, defined by $S_{\{\mathcal{I}\}} := \{f \in S \mid f(\mathcal{I}) = \mathcal{I}\}$. It can easily be seen that $S_{(\mathcal{I})}$ and $S_{(\mathcal{I}^\perp)}$ are normal subgroups in $S_{\{\mathcal{I}\}}$ and $S_{\{\mathcal{I}\}} = S$ if and only if $\mathcal{I}$ is a cardinal ideal. Also, it is not hard to see that the only non-trivial normal subgroups of S are the bounded permutation groups and the alternating subgroup generated by three cycles in S [1, 5]. If $\mathcal{I}$ is any set ideal, then the group $S_{(\mathcal{I})}$ is generated by involutions and each element of $S_{(\mathcal{I})}$ is a product of two involutions [4].

In the symmetric group S_n, for finite n except n=2,6, every automorphism is inner [10] and also every automorphism is inner for every infinite n [15]. In case of automorphism group, $\mathrm{Aut}(G)$ is complete for a non-abelian free group G [6]. For an infinite group, $\mathrm{Aut}(\mathbb{C})$, where $\mathbb{C}$ is the field of complex numbers, is complete by assuming CH [8], the group of measure-preserving transformations on [0,1] is complete [7], $\mathrm{Aut}(X)$ is complete for some complete Boolean algebra on X [14]. In

The author thank the University Grants Commission, New Delhi for providing the financial support.

S. B. Ramkumar (✉) · V. Renukadevi
Department of Mathematics, School of Mathematics and Computer Sciences, Central University of Tamil Nadu, Thiruvarur 610 005, India
e-mail: sbrkumar7@gmail.com

[12], assuming CH, it is shown that the stabilizer group of measure zero and category ideals on $\mathbb{R}$ is complete.

In 1999, Mishkin [13] proved a necessary and sufficient condition for the completeness of the stabilizer group $S_{\{\mathcal{I}\}}$ of a set ideal $\mathcal{I}$ in terms of Lagrange ideal by presuming that $\mathcal{I}$ is regular and top. Biryukov and Mishkin [3] found a necessary and sufficient condition for the stabilizer group $S_{\{\mathcal{I}\}}$ of a set ideal $\mathcal{I}$ to be complete in terms of reflexive ideals in 2002. Motivated by this, we attempt to find a relation connecting Lagrange and reflexive ideals and we try to improve the results. In [13], Mishkin proved some results on isomorphism between ideals by assuming that ideals are uniform(full). In this paper, we prove these results for non-uniform ideals, notably top ideals and we derive a relation connecting two set ideals and their stabilizer groups in Sect. 3.

2   Preliminaries

Let X be an infinite set and $\wp(X)$ be the set of all subsets of X. We recall that $\mathcal{I} \subseteq \wp(X)$ is said to be an *ideal* [11] if it satisfies the following two conditions: (a) If $A \in \mathcal{I}$ and $B \in \mathcal{I}$, then $A \cup B \in \mathcal{I}$ (finite additivity) and (b) If $A \in \mathcal{I}$ and $B \subset A$, then $B \in \mathcal{I}$ (heredity). The notation and basic definitions used in this paper are found in [2, 3, 13]. Let $\mathcal{I}^{+}$ (coideal) denote the complement of $\mathcal{I}$ in $\wp(X)$. If $A \subset X$, then $\wp(A)$ is an ideal on X. We call this ideal as principal ideal generated by A and is denoted by (A). Denote the non-proper ideal $(1) := \wp(X)$ and the trivial ideal $(0) := \{\emptyset\}$. An ideal $\mathcal{I}$ containing all singletons is said to be an *admissible* ideal on X.

For any ideal in a ring, naturally we define operations like sum, intersection, product, and quotient [17].

The sum $\mathcal{I} + \mathcal{J}$ of two ideals $\mathcal{I}$ and $\mathcal{J}$ is the ideal generated by $\mathcal{I} \cup \mathcal{J}$. If $\mathcal{I} + \mathcal{J} = (1)$, then $\mathcal{I}$ and $\mathcal{J}$ are said to be relatively prime (coprime). The ideal quotient of $\mathcal{I}$ by $\mathcal{J}$ (or colon ideal) is the ideal $(\mathcal{I} : \mathcal{J}) := \{A \subset X \mid (A) \cap \mathcal{J} \subset \mathcal{I}\}$.

Let κ, λ, ν denote infinite cardinals and ω stands for the first infinite cardinal and $\mathbf{c} = 2^{\omega}$ is the cardinality of the continuum. We use the standard notation, $[X]^{\lambda} = \{A \subset X \mid |A| = \lambda\}$ and $[X]^{<\lambda} = \{A \subset X \mid |A| < \lambda\}$. For each cardinal λ, the collection $[X]^{<\lambda}$ forms an ideal, which is referred as *cardinal* ideals on X.

Throughout this paper, $\mathcal{I}$ is assumed to be an admissible ideal on X. For a proper ideal $\mathcal{I}$ on X, we define the *norm* (or *uniformity*) $\|\mathcal{I}\| := \min \{|A| \mid A \in \mathcal{I}^{+}\}$ and the *height* $ht(\mathcal{I}) := \min \{\lambda \mid \mathcal{I} \subset [X]^{<\lambda}\}$. For the non-proper ideal, we put $\|(1)\| = ht((1)) = |X|^{+}$ where $|X|^{+}$ denotes the successor of $|X|$[9]. Clearly, $\|\mathcal{I}\| \leq ht(\mathcal{I})$ and the equality holds if and only if $\mathcal{I}$ is cardinal. An ideal $\mathcal{I}$ is said to be *full* (*uniform*) if $\|\mathcal{I}\| = |X|$, *confluent* if $ht(\mathcal{I}) = \|\mathcal{I}\|^{+}$ and *top* if $ht(\mathcal{I}) = |X|^{+}$. If $\mathcal{I}$ is top, then $\mathcal{I}$ contains a moiety of X (that is, a set $A \subset X$ with $|A| = |X \setminus A|$).

The λ-*section* of an ideal $\mathcal{I}$ is the ideal $\mathcal{I}|\lambda = \mathcal{I} \cap [X]^{<\lambda}$. The λ-*polar* of an ideal $\mathcal{I}$ on X is the ideal $\mathcal{I}^{\perp}(\lambda) := ([X]^{<\lambda} : \mathcal{I})$. $\mathcal{I}^{\perp} = \mathcal{I}^{\perp}(\|\mathcal{I}\|)$ is the polar of $\mathcal{I}$. Clearly, $\mathcal{I}^{\perp}(\lambda) = [X]^{<\lambda}$ if $\lambda < \|\mathcal{I}\|$ and $\mathcal{I}^{\perp}(\lambda) = (1)$ if and only if $\lambda \geq ht(\mathcal{I})$.

Let $\mathcal{I}$ be an ideal on X and let $\|\mathcal{I}\| \leq \lambda \leq ht(\mathcal{I})$. The λ-*adjoint* of $\mathcal{I}$ is the ideal $\mathcal{I}'(\lambda) := \{A \subset X \mid (A) \cap \mathcal{I} = [A]^{<\lambda}\}$. Thus, $\mathcal{I}'(\|I\|) = \mathcal{I}^{\perp}$. Clearly, $\mathcal{I} \cap \mathcal{I}'(\lambda) = \mathcal{I}|\lambda = \mathcal{I}'(\lambda)|\lambda$. Furthermore, $\|\mathcal{I}'(\lambda)\| = \|\mathcal{I}\|$.

Two ideals $\mathcal{I}$ on X and $\mathcal{J}$ on Y are said to be *isomorphic* ($\mathcal{I} \cong \mathcal{J}$) if there exists a bijection f from X to Y such that $f(\mathcal{I}) = \mathcal{J}$. An ideal $\mathcal{I}$ is called λ-*reflexive* if $\mathcal{I} \cong \mathcal{I}'(\lambda)|ht(\mathcal{I})$. If $\mathcal{I}$ is λ-reflexive for some $\lambda < ht(\mathcal{I})$, then it is said to be reflexive. An ideal $\mathcal{I}$ is said to be a *Lagrange* ideal if $\mathcal{I} \cong \mathcal{I}^{\perp}$. $\mathcal{I}$ is said to be *regular* if $\mathcal{I}^{\perp\perp} = \mathcal{I}$. It is clear that $\mathcal{I}^{\perp}$ is regular.

We recall that a centreless group G is *complete* if all automorphisms of G are inner. Let us denote Alt(X) and Sym(X) (or S), respectively, the alternating group and symmetric group on X. Let $f \in \text{Sym}(X)$. Then the *support* of f is supp$(f) :=$ $\{x \in X \mid f(x) \neq x\}$. Let $A \subseteq X$ and let S(A) = $\{f \in \text{Sym}(X) \mid \text{supp}(f) \subseteq A\}$ be the *pointwise stabilizer* of $X \smallsetminus A$. For an ideal $\mathcal{I}$, we define $S_{\langle \mathcal{I} \rangle} := \{f \in \text{Sym}(X) \mid \text{supp}(f) \in \mathcal{I}\}$. Denote bounded permutation groups $B_{\lambda} := S_{\langle [X]^{<\lambda} \rangle}$.

If there is no ambiguity of the group S, we denote the *normalizer* of $S_{\{\mathcal{I}\}}$ in S by $N(S_{\{\mathcal{I}\}})$ instead of $N_S(S_{\{\mathcal{I}\}})$. We denote Out$(S_{\{\mathcal{I}\}})$ as the set of all outer automorphisms of $S_{\{\mathcal{I}\}}$. The following lemmas will be useful in the sequel.

Lemma 1 [3, Lemma 3.2] *The symmetry groups of two ideals on X are isomorphic if and only if they are conjugate in Sym(X). Every automorphism of $S_{\{\mathcal{I}\}}$ is induced by an inner automorphism of its normalizer N in Sym(X). Thus, $Aut(S_{\{\mathcal{I}\}}) \cong N$.*

The following Remark 1 comes from Lemma 1,

Remark 1 $S_{\{\mathcal{I}\}}$ is complete if and only if it coincides with the normalizer $N(S_{\{\mathcal{I}\}})$ in Sym(X) [16].

Lemma 2 [3, Lemma 3.4] *Let $\mathcal{I}$ and $\mathcal{J}$ be two ideals on X with $\mathcal{J} \subset \mathcal{I}$ and $S_{\{\mathcal{I}\}} \leq S_{\{\mathcal{J}\}}$. Then $\mathcal{J} = \mathcal{I}|\kappa$ where $\kappa = ht(\mathcal{J})$.*

Lemma 3 [13, Lemma 3.2] *For every ideal $\mathcal{I}$ on X, the equality $S_{\{\mathcal{I}\}} = S_{\{\mathcal{I}^{\perp}\}}$ holds if and only if $\mathcal{I} = \mathcal{I}^{\perp\perp} \cap [X]^{<\alpha}$ where $\alpha = ht(\mathcal{I})$. Specifically, a top ideal $\mathcal{I}$ is regular if and only if $S_{\{\mathcal{I}\}} = S_{\{\mathcal{I}^{\perp}\}}$.*

Lemma 4 [3, Lemma 3.10] *Let $S_{\{\mathcal{I}\}} = S_{\{\mathcal{J}\}}$. Then $\mathcal{I} \cap \mathcal{J} = \mathcal{I}|\lambda = \mathcal{J}|\lambda$, where $\lambda = ht(\mathcal{I} \cap \mathcal{J})$. Furthermore, $\|\mathcal{I}\| = \|\mathcal{J}\| = \|\mathcal{I} \cap \mathcal{J}\|$.*

Theorem 1 [13, Theorem 3.1] *Let $\mathcal{I}$ be an ideal on X and let H be a non-trivial normal subgroup in $S_{\langle \mathcal{I} \rangle}$. Then either $H = Alt(X)$ or $H = S_{\langle \mathcal{I} \rangle} \cap B_{\lambda} = S_{\langle \mathcal{I} \cap [X]^{<\lambda} \rangle}$ where λ is the least cardinal with $H \subset B_{\lambda}$. Thus, the normal subgroup lattice of $S_{\langle \mathcal{I} \rangle}$ is a well ordered chain*

$$\{1\} \subset Alt(X) \subset B_{\omega} \subset \ldots \subset B_{\|I\|} \subset S_{\langle I \rangle} \cap B_{\|I\|^{+}} \subset \ldots \subset S_{\langle \mathcal{I} \rangle}.$$

3   Relation Between Lagrange and Reflexive Ideals

This section begins by showing the relationship between Lagrange and reflexive ideals. Later, we derive some results on isomorphic stabilizer groups of two ideals. The following Remark 2 shows that the cardinal ideals are neither reflexive nor Lagrange.

Remark 2 Cardinal ideal is neither reflexive nor Lagrange.

Proof Suppose that $\mathcal{I}$ is cardinal. Then $\mathcal{I} = [X]^{<\lambda}$ for some $\lambda \leq |X|$. It is clear that $ht(\mathcal{I}) = \|\mathcal{I}\| = \lambda$ and $\mathcal{I}^{\perp} = \{A \subseteq X \mid (A) \cap \mathcal{I} \subseteq [X]^{<\|\mathcal{I}\|}\} = \wp(X)$. This implies that $\mathcal{I}$ is not Lagrange, since $\mathcal{I}$ is proper and its polar is the non-proper ideal. Also, from the relation $\|\mathcal{I}\| \leq \lambda < ht(\mathcal{I})$, we see that no such λ exists and hence $\mathcal{I}$ is not reflexive.

The following Example 1 shows that a reflexive ideal $\mathcal{I}$ need not be a Lagrange ideal.

Example 1 Consider the real line $\mathbb{R}$. Partition $\mathbb{R}$ as $\mathbb{R} = A \cup B \cup C$, where $A = \mathbb{Q}$, $B = [\mathbb{R} \smallsetminus \mathbb{Q}] \cap (0, 1)$ and $C = [\mathbb{R} \smallsetminus (0, 1)] \cap [\mathbb{R} \smallsetminus \mathbb{Q}]$. Let $\mathcal{J}$ be a uniform(full) non-regular ideal on $\mathbb{Q}$. Then $\mathcal{J} \neq [\mathbb{Q}]^{<w}$. Indeed, if $\mathcal{J} = [\mathbb{Q}]^{<w}$, then $\mathcal{J}^{\perp} = \wp(\mathbb{Q})$ and $\mathcal{J}^{\perp\perp} = \mathcal{J}$ and hence $\mathcal{J}$ is regular, a contradiction. Define $\mathcal{I} := \mathcal{J} + [B]^{<\lambda} + (C)$, where $w < \lambda \leq |\mathbb{R}|$. Then it is easy to see that $\mathcal{I}$ is admissible and $\|\mathcal{I}\| = w$, because $\mathbb{Q} \notin \mathcal{I}$ and $\|\mathcal{I}\|$ is not finite. As $\|\mathcal{I}\| = w \neq |\mathbb{R}|$, $\mathcal{I}$ is not full. Also, $|C| = |\mathbb{R}|$ and $C \in \mathcal{I}$ implies that $ht(\mathcal{I}) = |\mathbb{R}|^{+}$ and hence $\mathcal{I}$ is top.

Now $w < \lambda \leq |\mathbb{R}|$ implies that $\lambda = |\mathbb{R}|$ and $\mathcal{I}'(|\mathbb{R}|) = \{A \subseteq \mathbb{R} \mid (A) \cap \mathcal{I} = [A]^{<|\mathbb{R}|}\}$. Since $\mathcal{J} \subseteq \wp(\mathbb{Q})$, $(\mathcal{J})^{<|\mathbb{R}|} = \wp(\mathcal{J})$ and $\mathcal{J} \cap \mathcal{I} = \mathcal{J}$, we have $\mathcal{J} \subseteq \mathcal{I}'(|\mathbb{R}|)$. Also, $[B]^{<|\mathbb{R}|} = \wp(B) \cap \mathcal{I}$ implies that $B \in \mathcal{I}'(|\mathbb{R}|)$ and $\wp(C) \cap \mathcal{I} = \wp(C) \neq [C]^{<|\mathbb{R}|}$. As $[C]^{<|\mathbb{R}|} \cap \mathcal{I} = [C]^{<|\mathbb{R}|}$, $[C]^{<|\mathbb{R}|} \subseteq \mathcal{I}'(|\mathbb{R}|)$. Since A, B and C partitioned $\mathbb{R}$, $\mathcal{I}'(|\mathbb{R}|) = \mathcal{J} + \wp(B) + [C]^{<|\mathbb{R}|}$. For any involution $f \in \text{Sym}(\mathbb{R})$ with $\text{supp}(f) = B \cup C$ interchanging B and C, we have $f(\mathcal{I}) = \mathcal{I}'(|\mathbb{R}|)$. Thus, $\mathcal{I}'(|\mathbb{R}|)|ht(\mathcal{I}) = \mathcal{I}'(|\mathbb{R}|) = f(\mathcal{I})$, because $\mathcal{I}$ is top. It follows that $\mathcal{I}$ is $|\mathbb{R}|$-reflexive and so $\mathcal{I}$ is reflexive.

We claim that $\mathcal{I}$ is not Lagrange. Suppose that $\mathcal{I}$ is a Lagrange ideal. Then $f(\mathcal{I}) = \mathcal{I}^{\perp}$ for some $f \in \text{Sym}(\mathbb{R})$. It is easy to see that A, B and C are not in $\mathcal{I}^{\perp}$, where $\mathcal{I}^{\perp} = \{D \subseteq \mathbb{R} \mid (D) \cap \mathcal{I} \subseteq [\mathbb{R}]^{<w}\}$. Now let U be any uncountable set in $\mathbb{R}$. If $|U \cap B|$ is not finite, then $U \notin \mathcal{I}^{\perp}$ and if $|U \cap B|$ is finite, then $U \cap C$ is infinite, since otherwise U should be countable, hence $U \notin \mathcal{I}^{\perp}$. Therefore, $\mathcal{I}^{\perp}$ does not contain any uncountable subset of $\mathbb{R}$. As $C \in \mathcal{I}$, $f(C) \in f(\mathcal{I}) = \mathcal{I}^{\perp}$, a contradiction. Thus, $f(\mathcal{I}) \neq \mathcal{I}^{\perp}$ for all $f \in \text{Sym}(\mathbb{R})$.

The following Example 2 gives an ideal which is both reflexive and Lagrange.

Example 2 Consider the real line $\mathbb{R}$ with usual ordering '$\leq$', that is, $(\mathbb{R}, \leq)$ is an infinite linearly ordered set. It is easy to see that $(\mathbb{R}, \leq)$ is isomorphic to $(\mathbb{R}, \geq)$ by means of the function $f(x) = -x$. Let $\mathcal{I}$ be the ideal of all well-ordered subsets in $\mathbb{R}$. As every finite set is well ordered and $\mathbb{Q}$ is not well ordered, $\|\mathcal{I}\| = \omega$ and since $\|\mathcal{I}\| = \omega \neq |\mathbb{R}|$, $\mathcal{I}$ is not full. Since $\mathcal{I}$ does not contain any uncountable set, $ht(\mathcal{I}) = |\mathbb{R}|$ and $\mathcal{I}$ is not top. From the relation $\|\mathcal{I}\| \leq \lambda < ht(\mathcal{I})$, we have $\lambda = \|\mathcal{I}\| = \omega$ and $\mathcal{I}'(\lambda) = \mathcal{I}'(\|\mathcal{I}\|) = \mathcal{I}^\perp$. We can easily prove that $\mathcal{I}^\perp$ consists of all well ordered subsets in $(\mathbb{R}, \geq)$ and $\mathcal{I}$ is Lagrange. Now, $\mathcal{I}'(\lambda)|ht(\mathcal{I}) = \mathcal{I}'(\|\mathcal{I}\|)|ht(\mathcal{I}) = \mathcal{I}^\perp$, because $ht(\mathcal{I}) = ht(\mathcal{I}^\perp)$. It follows that $\mathcal{I}$ is $\|\mathcal{I}\|$-reflexive and hence reflexive.

The following Example 3 shows the existence of an ideal which is neither reflexive nor Lagrange.

Example 3 Consider the ideal $\mathcal{I} = [\mathbb{Q}]^{<|\mathbb{Q}|} \cup (B)$ where $B = \mathbb{R} \setminus \mathbb{Q}$. Then $\|\mathcal{I}\| = \omega$, since $\mathbb{Q} \notin \mathcal{I}$. As $\|\mathcal{I}\| = \omega \neq |\mathbb{R}|$, $\mathcal{I}$ is not full. Since $|B| = |\mathbb{R} \setminus \mathbb{Q}| = |\mathbb{R}|$ and $B \in \mathcal{I}$, $\mathcal{I}$ is top. We will find $\mathcal{I}^\perp$. If A is a finite subset of $\mathbb{R}$, then $A \in \mathcal{I}$ and $(A) \cap \mathcal{I} = (A) \subseteq [\mathbb{R}]^{<\omega}$ and hence $A \in \mathcal{I}^\perp$. Let A be a countable subset of $\mathbb{R}$. If $A \cap \mathbb{Q} = \emptyset$, then $A \in \mathcal{I}$ and $(A) \cap \mathcal{I} = (A) \not\subseteq [\mathbb{R}]^{<\omega}$. If $A \cap \mathbb{Q} \neq \emptyset$, then $(A) \cap \mathcal{I} \subseteq [\mathbb{R}]^{<\omega}$ if $A \cap B$ is finite and $A \notin \mathcal{I}^\perp$, if $A \cap B$ is infinite. Let A be an uncountable set. Then $A \cap B$ is uncountable, otherwise A would be countable. Also, $(A) \cap (B) \subseteq \mathcal{I}$ implies that $(A) \cap (B) \cap \mathcal{I} \subseteq (A) \not\subseteq [\mathbb{R}]^{<\omega}$, hence $A \notin \mathcal{I}^\perp$. Therefore, $\mathcal{I}^\perp = (\mathbb{Q}) \cup [\mathbb{R} \setminus \mathbb{Q}]^{<\omega}$.

We now claim that $\mathcal{I}$ is not Lagrange. Suppose that $\mathcal{I} \cong \mathcal{I}^\perp$. Then $f(\mathcal{I}) = \mathcal{I}^\perp$ for some $f \in \text{Sym}(\mathbb{R})$. If A is any uncountable subset of $\mathbb{R} \setminus \mathbb{Q}$, then $A \in \mathcal{I}$ and so $f(A) \in f(\mathcal{I}) = \mathcal{I}^\perp$. Since A is uncountable and f is a bijection, $f(A)$ is an uncountable set in $\mathcal{I}^\perp$, which is a contradiction. Hence $f(\mathcal{I}) \neq \mathcal{I}^\perp$ for every $f \in \text{Sym}(\mathbb{R})$. That is, $\mathcal{I}$ is not Lagrange.

We show that $\mathcal{I}$ is not reflexive. Let λ such that $\|\mathcal{I}\| \leq \lambda \leq ht(\mathcal{I})$. Then $\omega \leq \lambda < |\mathbb{R}|^+$. This implies that either $\lambda = \omega$ or $\lambda = |\mathbb{R}|$. If $\lambda = \omega$, then $\mathcal{I}'(\lambda) = \mathcal{I}'(\|\mathcal{I}\|) = \mathcal{I}^\perp$ and $\mathcal{I}'(\lambda)|ht(\mathcal{I}) = \mathcal{I}'(\lambda) \cap [\mathbb{R}]^{<ht(\mathcal{I})} = \mathcal{I}^\perp \cap [\mathbb{R}]^{<|\mathbb{R}|^+} = \mathcal{I}^\perp$. If $\mathcal{I} \cong \mathcal{I}'(\|\mathcal{I}\|)|ht(\mathcal{I})$, then $\mathcal{I} \cong \mathcal{I}^\perp$, a contradiction. Thus, $\mathcal{I}$ is not $\|\mathcal{I}\|$-reflexive. Suppose that $\lambda = |\mathbb{R}|$. Then $\mathcal{I}'(|\mathbb{R}|) = \{A \subseteq \mathbb{R} \mid (A) \cap \mathcal{I} = [A]^{<|\mathbb{R}|}\}$. Let A be any finite subset of $\mathbb{R}$. Then $[A]^{<|\mathbb{R}|} = \wp(A)$ and $(A) \cap \mathcal{I} = (A)$, hence $A \in \mathcal{I}'(|\mathbb{R}|)$. If A is any countable set in $\mathbb{R}$, then $[A]^{<|\mathbb{R}|} = \wp(A)$. If $A \subseteq \mathbb{Q}$, then $(A) \cap \mathcal{I} \neq \wp(A)$, since $A \notin \mathcal{I}$. This implies that $A \notin \mathcal{I}'(|\mathbb{R}|)$. If $A \subseteq \mathbb{R} \setminus \mathbb{Q}$, then $(A) \cap \mathcal{I} = \wp(A)$ and so $A \in \mathcal{I}'(|\mathbb{R}|)$. If $A \cap \mathbb{Q} \neq \emptyset$ and $A \cap [\mathbb{R} \setminus \mathbb{Q}] \neq \emptyset$, then $A \notin \mathcal{I}'(|\mathbb{R}|)$ if $A \cap \mathbb{Q}$ is infinite and $A \in \mathcal{I}'(|\mathbb{R}|)$ if $A \cap \mathbb{Q}$ is finite. Let A be any uncountable set. Then $A \cap [\mathbb{R} \setminus \mathbb{Q}]$ is uncountable and $A \cap [\mathbb{R} \setminus \mathbb{Q}] \in \mathcal{I}$. If $A \in \mathcal{I}'(|\mathbb{R}|)$, then $B = A \cap [\mathbb{R} \setminus \mathbb{Q}] \in \mathcal{I}'(\mathbb{R})$. But $(B) \cap \mathcal{I} = (B) \neq [B]^{<|\mathbb{R}|}$, since B is uncountable and $B \notin [B]^{<|\mathbb{R}|}$, a contradiction. Hence $A \notin \mathcal{I}'(|\mathbb{R}|)$. Thus, $\mathcal{I}'(|\mathbb{R}|) = [\mathbb{R}]^{<\omega} \cup [\mathbb{R} \setminus \mathbb{Q}]^{<|\mathbb{R}|}$.

Suppose $\mathcal{I}$ is $|\mathbb{R}|$-reflexive. Then $\mathcal{I} \cong \mathcal{I}'(|\mathbb{R}|)|ht(\mathcal{I})$. As $ht(\mathcal{I}) = |\mathbb{R}|^+$, $\mathcal{I}'(|\mathbb{R}|) \cap [\mathbb{R}]^{<|\mathbb{R}|^+} = \mathcal{I}'(|\mathbb{R}|)$. Therefore, $f(\mathcal{I}) = \mathcal{I}'(|\mathbb{R}|)$ for some $f \in \text{Sym}(\mathbb{R})$. As $\mathbb{R} \setminus \mathbb{Q} \in \mathcal{I}$, we have $f(\mathbb{R} \setminus \mathbb{Q}) \in f(\mathcal{I}) \in \mathcal{I}'(|\mathbb{R}|)$. This implies that $\mathcal{I}'(|\mathbb{R}|)$ contains an uncountable set, which is a contradiction. Hence $\mathcal{I} \neq \mathcal{I}'(|\mathbb{R}|)|ht(\mathcal{I})$. Thus, $\mathcal{I}$ is not reflexive.

Theorem 2 *If $\mathcal{I}$ is a Lagrange ideal, then $\mathcal{I}$ is reflexive. The converse holds if $\mathcal{I}$ is full.*

Proof From $\mathcal{I} \cong \mathcal{I}^{\perp}$, $\mathcal{I}'(\|\mathcal{I}\|) = \mathcal{I}^{\perp}$ and $ht(\mathcal{I}) = ht(\mathcal{I}^{\perp})$, we have $\mathcal{I}$ is $\|\mathcal{I}\|$-reflexive and hence $\mathcal{I}$ is reflexive.

Conversely, suppose that $\mathcal{I}$ is full. Then $\|\mathcal{I}\| = |X|$ and $ht(\mathcal{I}) = |X|^{+}$. From the relation $\|\mathcal{I}\| \leq \lambda < ht(\mathcal{I})$, we have $\lambda = \|\mathcal{I}\|$. This implies that $\mathcal{I}$ is $\|\mathcal{I}\|$-reflexive and $\mathcal{I} \cong \mathcal{I}^{\perp}$. Thus, $\mathcal{I}$ is Lagrange.

In [13, Theorem 3.6], Mishkin proved that if $\mathcal{I}$ is a regular ideal and $\mathcal{J}$ is a top ideal on an infinite set X and $S_{\{\mathcal{I}\}} = S_{\{\mathcal{J}\}}$ (resp. $S_{\{\mathcal{I}\}} \cong S_{\{\mathcal{J}\}}$), then either $\mathcal{J} = \mathcal{I}$ or $\mathcal{J} = \mathcal{I}^{\perp}$ (resp. either $\mathcal{J} \cong \mathcal{I}$ or $\mathcal{J} \cong \mathcal{I}^{\perp}$). In Theorems 3 and 4 below, we prove this result by assuming the condition that the ideals $\mathcal{J}$ and $\mathcal{I} \cap \mathcal{J}$ are top. Example 4 shows that the above condition does not imply that $\mathcal{I}$ is regular.

Theorem 3 *If the stabilizer group of a top ideal $\mathcal{J}$ on X is equal to that of an ideal $\mathcal{I}$, then either $\mathcal{J} = \mathcal{I}$ or $\mathcal{J} = \mathcal{I}^{\perp}$ provided $\mathcal{I} \cap \mathcal{J}$ is top whenever $\mathcal{J} \nsubseteq \mathcal{I}^{\perp}$.*

Proof Suppose that $S_{\{\mathcal{I}\}} = S_{\{\mathcal{J}\}}$. If $\mathcal{J} \subseteq \mathcal{I}$, then by Lemma 2, $\mathcal{J} = \mathcal{I}|\kappa$ where $\kappa = ht(\mathcal{J})$. Since $\mathcal{J}$ is top, $ht(\mathcal{J}) = |X|^{+}$ and hence $\mathcal{J} = \mathcal{I}$. Suppose that $\mathcal{J} \nsubseteq \mathcal{I}$. If $\mathcal{J} \subseteq \mathcal{I}^{\perp}$, then $S_{(\mathcal{J})} \leq S_{(\mathcal{I}^{\perp})}$ and it follows that $S_{(\mathcal{J})}$ is normal in $S_{(\mathcal{I}^{\perp})}$. It follows from Theorem 1 that $\mathcal{J} = \mathcal{I}^{\perp}$, because $\mathcal{J}$ is top. Finally, if $\mathcal{J} \nsubseteq \mathcal{I}^{\perp}$, then as $\mathcal{I} \cap \mathcal{J}$ is top, then by Lemma 4, $\mathcal{I} \cap \mathcal{J} = \mathcal{I}$, a contradiction. Hence $\mathcal{J} \nsubseteq \mathcal{I}^{\perp}$ is not possible and it completes the proof.

Theorem 4 *If the stabilizer group of a top ideal $\mathcal{J}$ on X is isomorphic to that of an ideal $\mathcal{I}$, then either $\mathcal{J} \cong \mathcal{I}$ or $\mathcal{J} \cong \mathcal{I}^{\perp}$ if $\mathcal{I} \cap f(\mathcal{J})$ is top whenever $f(\mathcal{J}) \nsubseteq \mathcal{I}^{\perp}$ for every $f \in Sym(X)$.*

Example 4 Let $\mathcal{I}_1$ and $\mathcal{I}_2$ respectively denote the set of all measure zero subsets of $[0, 2]$ and the set of all measure zero subsets of $[1, 3]$. Define $\mathcal{I} := \{A \subset \mathbb{R} \mid A$ is countable or $A \in \mathcal{I}_1\}$ and $\mathcal{J} := \{B \subset \mathbb{R} \mid B$ is countable or $B \in \mathcal{I}_2\}$. Then $\mathcal{J} \nsubseteq \mathcal{I}^{\perp}$, $\mathcal{I} \cap \mathcal{J}$ is top but $\mathcal{I}$ is not regular.

Remark 3 In Theorems 3 and 4, the condition that $\mathcal{J}$ is a top ideal cannot be dropped from the hypothesis. For example, consider $\mathcal{I}_f$, the collection of all finite subsets and $\mathcal{I}_c$, the collection of all countable subsets on any uncountable set X. These ideals are distinct, not top and one is not equal to the polar of other but both ideals have the same stabilizer group.

The converse of Theorems 3 and 4 holds only if $\mathcal{I}$ is regular top as in the following Corollary 1.

Corollary 1 *Let $\mathcal{I}$ be a regular top ideal and $\mathcal{J}$ be a top ideal on X. If either $\mathcal{J} = \mathcal{I}$ or $\mathcal{J} = \mathcal{I}^{\perp}$, then $S_{\{\mathcal{I}\}} = S_{\{\mathcal{J}\}}$.*

Proof If $\mathcal{I} = \mathcal{J}$ (resp. $\mathcal{I} \cong \mathcal{J}$), then there is nothing to prove. Assume that $\mathcal{J} = \mathcal{I}^{\perp}$ (resp. $\mathcal{J} \cong \mathcal{I}^{\perp}$). It follows from Lemma 3 that $S_{\{\mathcal{I}\}} = S_{\{\mathcal{J}\}}$.

Remark 4 In Corollary 1, the condition that $\mathcal{I}$ is a top ideal cannot be dropped from the hypotheis. Because if $\mathcal{I}$ is not top, then $\mathcal{I}$ does not contain any moiety of X and since $\mathcal{J}$ is top, we can find moieties A and B of X such that $A \in \mathcal{J}$ and $B \notin \mathcal{J}$. Now we can find a bijection $f \in S$ such that $f(A) = B$. It follows that $f \in S_{\{\mathcal{I}\}}$, $f \notin S_{\{\mathcal{J}\}}$ and hence $S_{\{\mathcal{I}\}} \neq S_{\{\mathcal{J}\}}$ (resp. $S_{\{\mathcal{I}\}} \not\cong S_{\{\mathcal{J}\}}$). Thus, to prove the result, the assumption that $\mathcal{I}$ is a top ideal is mandatory.

Corollary 2 [13] *If $\mathcal{I}$ and $\mathcal{J}$ are uniform set ideals on X and $S_{\{\mathcal{I}\}} = S_{\{\mathcal{J}\}}$ (resp. $S_{\{\mathcal{I}\}} \cong S_{\{\mathcal{J}\}}$), then either $\mathcal{J} = \mathcal{I}$ or $\mathcal{J} = \mathcal{I}^{\perp}$ (resp. either $\mathcal{J} \cong \mathcal{I}$ or $\mathcal{J} \cong \mathcal{I}^{\perp}$).*

In [13, Theorem 3.7], Mishkin proved that when $\mathcal{I}$ is full or regular top ideal on an infinite set, the stabilizer group $S_{\{\mathcal{I}\}}$ is complete if and only if $\mathcal{I}$ is not Lagrange. Also, in [3, Theorem 3.19], it was proved that the stabilizer group $S_{\{\mathcal{I}\}}$ is complete if and only if $\mathcal{I}$ is non-reflexive. Example 1 shows that the condition, $\mathcal{I}$ is regular, cannot be dropped from the hypothesis. This is because all Lagrange ideals are reflexive.

Theorem 5 [13, Theorem 3.7] *Suppose that $\mathcal{I}$ is a top ideal on an infinite set X. If the stabilizer group $S_{\{\mathcal{I}\}}$ is complete, then $\mathcal{I}$ is not a Lagrange ideal.*

Example 1 shows that the converse of the above theorem need not be true.

Theorem 6 *If $\mathcal{I}$ is a Lagrange ideal, then the outer automorphism group of $S_{\{\mathcal{I}\}}$ is a cyclic group of order at most 2.*

Proof If possible, choose $f \in N(S_{\{\mathcal{I}\}}) \setminus S_{\{\mathcal{I}\}}$, otherwise $N(S_{\{\mathcal{I}\}}) = S_{\{\mathcal{I}\}}$. Then $f(\mathcal{I}) \neq \mathcal{I}$. If $f(\mathcal{I}) \not\subseteq \mathcal{I}^{\perp}$, then we can prove easily that $f(\mathcal{I}) \subseteq \mathcal{I}^{\perp\perp} = \mathcal{I}$. It follows from Lemma 2 that $f(\mathcal{I}) = \mathcal{I}$, a contradiction. Hence $f(\mathcal{I}) \subset \mathcal{I}^{\perp}$ and by Lemma 2, $f(\mathcal{I}) = \mathcal{I}^{\perp}$. Now $f^2(\mathcal{I}) = f(\mathcal{I}^{\perp}) = \mathcal{I}$. Thus, $f^2 \in S_{\{\mathcal{I}\}}$ and $\mathrm{Out}(S_{\{\mathcal{I}\}}) \cong N(S_{\{\mathcal{I}\}})/S_{\{\mathcal{I}\}} \cong \mathbb{Z}_2$.

4 Conclusion

In this article, we found a necessary and sufficient condition for the isomorphism between stabilizer groups of two set ideals on an infinite set. Also, we studied the completeness of the stabilizer group of set ideals. In the future, one can explore the completeness of stabilizer group and the automorphism group of the stabilizer group of different set ideals. Through this, one can investigate the Lie structure for stabilizer group.

Acknowledgements The authors sincerely thank the anonymous referee for the valuable suggestions.

References

1. Baer R (1934) Die Kompositionsreihe der Gruppe aller eineindeutigen Abbildungen einer unendlichen Menge auf sich. Studia Math 5:15–17
2. Biryukov P, Mishkin V (2000) Set ideals with maximal symmetry group and minimal dynamical systems. Bull London Math Soc 32(1):39–46
3. Biryukov P, Mishkin V (2002) Set ideals with isomorphic symmetry groups. Bull London Math Soc 34(1):37–45
4. Dixon JD, Mortimer B (1966) Permutation groups. Springer-Verlag, Berlin
5. Drost M, Göbel R (1979) On a theorem of Baer, Schreier, and Ulam for permutations. J Algebra 58:282–290
6. Dyer J, Formanek E (1975) The automorphism group of a free group is complete. J London Math Soc 11(2):181–190
7. Eigen S (1982) The group of measure preserving transformations of [0,1] has no outer automorphisms. Math Ann 259:259–270
8. Evans DM, Laskar D (1997) The automorphism group of field of complex numbers is complete. In: Evans DM (ed) Model theory of groups and automorphism groups. London Math Soc Lecture Note Ser, vol 244. Cambridge University Press, pp 115–125 (1997)
9. Halmos Paul R (1974) Naive set theory. Springer, New York
10. Hölder, O. (1895) Bildung zusammengesetzter Gruppen. Math Ann 46:321–422
11. Kuratowski K (1972) Introduction to set theory and topology, 2nd edn. Pergamon Press Ltd., New York
12. Mishkin V (1989) Autohomeomorphism groups of spaces with unique non-isolated point. Comment Math Univ Carolinae 30(1):89–94
13. Mishkin V (1999) Set ideals with complete symmetry group and partition ideals. Bull London Math Soc 31(6):649–660
14. Rubin M (1989) On the reconstruction of Boolean algebras from their automorphism groups. In: Monk JD (ed) Handbook of Boolean algebras. Elsevier, pp 547–606
15. Schreier J, Ulam S (1937) Über die Automprphismen der permutation-Gruppe der naturlichen Zahlenfolge. Fund Math 28(1):258–260
16. Scott WR (1964) Group theory. Prentice-Hall, Englewood Gliffs, New York
17. Van der waerden BL (1967) Algebra II, 5th edn. Springer-Verlag, Berlin, Heidelberg

λ—Statistical Derivative

Samar Idris and Rifat Çolak

Abstract This investigation article concentrates on de ning λ—statistical derivative, (V, λ)—statistical derivative, and strong (V, λ)—statistical derivative using statistical and Ces ro derivatives. Established on the rst concept, we examine the a nity between statistical and ordinary derivatives. Furthermore, we de ned (V, λ)—continuity and studied the relationship between it and (V, λ)—derivative. The connection between λ—statistical derivative and strong (V, λ)—statistical derivative has been discussed too. Finally, we examined the a nity between strong. (V, μ)—statistical derivative and strong (V, λ)—statistical derivative, under some conditions, for any λ, μ in Λ.

Keywords Statistical derivative · λ—statistical derivative · (V, λ)—statistical derivative · Strong (V, λ)—statistical derivative

1 Introduction

The concept of statistical convergence was instructed independently by Fast [1] and studied by multifarious authors [2–4]. Statistical convergence is a kind of convergence of real sequences more powerless than ordinary convergence. It is represented as follows: A sequence $x = (x_k)$ is said to be statistically convergent to the number L if for every

$$\lim_{n \to \infty} \frac{1}{n} |\{k \le n : |x_k - L| \ge \varepsilon\}| = 0$$

S. Idris (✉)
Graduate School of Sciences, Firat University, Elazığ, Turkey
e-mail: samar.idris23@gmail.com
URL: http://www.rat.edu.tr

R. Çolak
Department of Mathematics, Firat University, Elazığ, Turkey
URL: http://www.rat.edu.tr

D. Giri et al. (eds.), *Proceedings of the Tenth International Conference on Mathematics and Computing*, Lecture Notes in Networks and Systems 964,
https://doi.org/10.1007/978-981-97-2066-8_17

183

where the vertical bars indicate the number of elements in the enclosed set, in this case, we write S—$\lim x = L$ or $x_k \to L(S)$, and S denotes the set of all statistically convergent sequences.

Likewise, Mursaleen introduced the Concept of λ-statistical convergence to generalize the notion of statistical convergence [5].

Let $\lambda = (\lambda_n)$ be a non-decreasing sequence of positive numbers tending to ∞ such that $\lambda_{n+1} \leq \lambda_n + 1$, $\lambda_1 = 1$. We write Λ to denote the set of such sequences, that is

$$\Lambda = \{\lambda = (\lambda_n) : \lambda_{n+1} \leq \lambda_n + 1, \lambda_1 = 1, \lambda_n \to \infty (n \to \infty), \lambda_n > 0 \quad \text{for all n}\}.$$

A sequence $x = (x_k)$ is said to be λ—statistically convergent to the number L if for every. $\varepsilon > 0$

$$\lim_{n \to \infty} \frac{1}{\lambda_n} |\{k \in I_n : |x_k - L| \geq \varepsilon\}| = 0$$

where $I_n = [n - \lambda_n + 1, n]$ [5] and in this case, we write S_λ—$\lim x = L$ or $x_k \to L(S_\lambda)$. The set of all λ—statistically convergent sequences is denoted by S_λ. λ—statistical convergence has numerous attractive properties. One of them is that every statistically convergent sequence is also λ—statistically convergent for every λ in Λ.

The generalized de la ValØe-Pousin mean of a sequence (x_k) is de ned by

$$t_n(x) = \frac{1}{\lambda_n} \sum_{k \in I_n} x_k.$$

A sequence $x = (x_k)$ is said to be (V, λ)—summable to a number L if

$$k_n(x) \to L(n \to \infty).$$

If $\lambda_n = n$, then (V, λ)—summability reduces to $(C, 1)$—summability. We write

$$[C, 1] = \left\{ x = (x_k) : \exists L \in \mathbb{R}, \lim_{n \to \infty} \frac{1}{n} \sum_{k=1}^{n} |x_k - L| = 0 \right\}$$

and

$$[V, \lambda] = \left\{ x = (x_k) : \exists L \in \mathbb{R}, \lim_{n \to \infty} \frac{1}{\lambda_n} \sum_{k \in I_n} |x_k - L| = 0 \right\}$$

for the sets of sequences $x = (x_k)$ which are strongly Ces ro summable and strongly (V, λ)—summable to L, i.e., $x_k \to L[C, 1]$ and $x_k \to L[V, \lambda]$, respectively. (V, λ)—summability is a subject of busy mathematics research, including di erential equations, in nite products, and Fourier series.

2 Main Results

Nuray [6] has recently introduced the idea of Ces ro derivative and statistical deriva-
tive and studied some basic properties and signi cant connections relations between
Ces ro and statistical derivatives. It is represented as follows:
   A function $f: \text{IR} \rightarrow \text{IR}$ has a Ces ro derivative $w \in \text{IR}$ at a point $x_0 \in \text{IR}$ if

$$\lim_{n \to \infty} \frac{1}{n} \sum_{k=1}^{n} \frac{f(x_k + x_0) - f(x_0)}{x_k} = w$$

holds whenever $x_n > 0$ and $\lim_{n \to \infty} x_n = 0$.
   A function $f: \text{IR} \rightarrow \text{IR}$ has a statistical derivative $w \in \text{IR}$ at a point $x_0 \in \text{IR}$ if

$$\lim_{n \to \infty} \frac{1}{n} \left| \left\{ k \leq n : \left| \frac{f(x_k + x_0) - f(x_0)}{x_k} - w \right| \geq \varepsilon \right\} \right| = 0$$

holds for every $\varepsilon > 0$ whenever $x_n > 0$ and $\lim_{n \to \infty} x_n = 0$. If the statistical derivative
of the function f is w at the point $x_0 \in \text{IR}$, then we write $st{-}f^0(x_0) = w$.
   Nuray's appointment on Ces ro and statistical derivatives is unique and impressive,
opening up the opportunity for di erent studies in this area. For instance, studying the
connection between these and other derivatives, such as the ordinary and fractional
derivatives, would be fascinating. It would also be impressive to study the applications of these derivatives to real-world problems. Established on this idea, we begin
studying the a nity between statistical and ordinary derivatives.

Theorem 1 If a function has the derivative at a point, then, it also has a statistical
derivative at that point, but the converse may not be correct. In another term if $f^0(x_0)$
exists then $st{-}f^0(x_0) = f^0(x_0)$.

Proof. Proof of the rst part is easy. For the converse, we provide the next example.

Example. Let us consider the function $f: \text{IR} \rightarrow \text{IR}$ de ned by $f(x) = [x]$, where $[x]$
is the greatest integer function. Let us consider any sequence (x_n) such that $x_n > 0$
for every n and $\lim_{n \to \infty} x_n = 0$. Then, given any $\varepsilon > 0$ ($\varepsilon < 1$) there exists a positive
integer n_0 such that

$$[x_n] \leq x_n = |x_n - 0| = |x_n| < \varepsilon$$

for every $n \geq n_0$. Now, given any $\varepsilon > 0$ and for every $n \geq n_0$ we may write

$$\frac{1}{n}\left|\left\{k \le n : \left|\frac{f(x_k + 0) - f(0)}{x_k} - 0\right| \ge \varepsilon\right\}\right|$$

$$= \frac{1}{n}\left|\left\{k \le n : \left|\frac{[x_k + 0] - [0]}{x_k}\right| \ge \varepsilon\right\}\right|$$

$$= \frac{1}{n}\left|\left\{k \le n : \left|\frac{[x_k]}{x_k}\right| \ge \varepsilon\right\}\right| \le \frac{n_0}{n}$$

Therefore

$$\lim_{n \to \infty} \frac{1}{n}\left|\left\{k \le n : \left|\frac{f(x_k + 0) - f(0)}{x_k} - 0\right| \ge \varepsilon\right\}\right| \le \lim_{n \to \infty} \frac{n_0}{n} = 0$$

Hence, f has a statistical derivative at 0, but on the other hand, it is well known that f has no derivative at 0.tu.

Behind that, we continue with some belongings.

Theorem 2 Let the functions f, g: IR $\to$ IR be given. Suppose both f and g have statistical derivatives $w_1, w_2 \in$ IR at a point $x_0 \in$ IR, respectively. Then, we have the following facts.

(i) The function $f + g$ has a statistical derivative at the point $x_0 \in$ IR and $st\!-\!(f + g)0(x_0) = st - f0(x_0) + st - g0(x_0)$

(ii) For a constant c, the function cf has a statistical derivative at the point $x_0 \in$ IR and $st - (c.f)0(x_0) = c.st - f0(x_0)$

Proof. (*i*) Suppose that the functions f and g have statistical derivatives at the point $x_0 \in$ IR and $st\!-\!f0(x_0) = w_1$, $st\!-\!g0(x_0) = w_2$. Then,

$$\lim_{n \to \infty} \frac{1}{n}\left|\left\{k \le n : \left|\frac{f(x_k + x_0) - f(x_0)}{x_k} - w_1\right| \ge \varepsilon\right\}\right| = 0$$

and

$$\lim_{n \to \infty} \frac{1}{n}\left|\left\{k \le n : \left|\frac{g(x_k + x_0) - g(x_0)}{x_k} - w_2\right| \ge \varepsilon\right\}\right| = 0$$

for every $\varepsilon > 0$. Now, we may write

$$\lim_{n \to \infty} \frac{1}{n} \left| \left\{ k \le n : \left| \frac{(f+g)(x_k + x_0) - (f+g)(x_0)}{x_k} - (w_1 + w_2) \right| \ge \varepsilon \right\} \right|$$

$$= \lim_{n \to \infty} \frac{1}{n} \left| \left\{ k \le n : \left| \frac{(f(x_k + x_0) + g(x_k + x_0)) - (f(x_0) + g(x_0))}{x_k} - (w_1 + w_2) \right| \ge \varepsilon \right\} \right|$$

$$= \lim_{n \to \infty} \frac{1}{n} \left| \left\{ k \le n : \left| \left(\frac{f(x_k + x_0) + f(x_0)}{x_k} - w_1 \right) + \left(\frac{g(x_k + x_0) + g(x_0)}{x_k} - w_2 \right) \right| \ge \varepsilon \right\} \right|$$

$$\le \lim_{n \to \infty} \frac{1}{n} \left| \left\{ k \le n : \left| \frac{f(x_k + x_0) - f(x_0)}{x_k} - w_1 \right| \ge \frac{\varepsilon}{2} \right\} \right|$$

$$+ \lim_{n \to \infty} \frac{1}{n} \left| \left\{ k \le n : \left| \frac{g(x_k + x_0) - g(x_0)}{x_k} - w_2 \right| \ge \frac{\varepsilon}{2} \right\} \right|$$

and taking limit as $n \to \infty$ we get that

$$\lim_{n \to \infty} \frac{1}{n} \left| \left\{ k \le n : \left| \frac{(f+g)(x_k + x_0) - (f+g)(x_0)}{x_k} - (w_1 + w_2) \right| \ge \varepsilon \right\} \right| = 0$$

and this means that st—$(f+g)0(x_0) = st$—$f0(x_0) + st$—$g0(x_0)$.

The proof of (ii) is similar, so we skip it.

Defnition 1 Let $\lambda \in \Lambda$ be given. A function $f \colon \mathrm{IR} \to \mathrm{IR}$ has the λ—statistical derivative $w \in \mathrm{IR}$ at a point $x_0 \in \mathrm{IR}$ if

$$\lim_{n \to \infty} \frac{1}{\lambda_n} \left| \left\{ k \in I_n : \left| \frac{f(x_k + x_0) - f(x_0)}{x_k} - w \right| \ge \varepsilon \right\} \right| = 0$$

holds for every $\varepsilon > 0$, whenever $x_n > 0$ and $\lim_{n \to \infty} X_n = 0$.

A matching de nition to this De nition is as follows.

A function $f \colon \mathrm{IR} \to \mathrm{IR}$ has a λ—statistical derivative $w \in \mathrm{IR}$ at a point $x_0 \in \mathrm{IR}$ if

$$\lim_{n \to \infty} \frac{1}{\lambda_n} \left| \left\{ k \in I_n : \left| \frac{f(x_k + x_0) - f(x_0 - x_k)}{2x_k} - w \right| \ge \varepsilon \right\} \right| = 0$$

holds for every $\varepsilon > 0$, whenever $x_n > 0$ and $\lim_{n \to \infty} X_n = 0$.

Remark 1. Note that if we take $\lambda_n = n$, the overhead de nition coincides with the de nition of the statistical derivative. So, from that point of view, λ—statistical derivative is an extension of the statistical derivative.

Theorem 3 Let a function $f \colon \mathrm{IR} \to \mathrm{IR}$ be given and $\lambda = (\lambda_n)$, $\mu = (\mu_n)$ be two sequences in Λ such that $\lambda_n \le \mu_n$ for all $n \in \mathrm{IN}$. Then, we have the next facts.

(i) If f has μ—statistical derivative, then f has λ—statistical derivative in the case

$$\liminf_{n \to \infty} \frac{\lambda_n}{\mu_n} > 0 \tag{1}$$

(ii) If f has λ—statistical derivative, then f has μ—statistical derivative in the case

$$\liminf_{n\to\infty} \frac{\lambda_n}{\mu_n} > 1 \tag{2}$$

Proof. (i) Suppose the function $f: \mathrm{IR} \to \mathrm{IR}$ has μ—statistical derivative at a point $x_0 \in \mathrm{IR}$ and let $\lambda_n \le \mu_n$ for all $n \in \mathrm{IN}$, so $I_n \subset J_n$, where $I_n = [n{-}\lambda_n + 1, n]$ and $J_n = [n{-}\mu_n + 1, n]$. Now, for any $\varepsilon > 0$, we can write

$$\left\{ k \in J_n : \left| \frac{f(x_k + x_0) - f(x_0)}{x_k} - w \right| \ge \varepsilon \right\} \supset \left\{ k \in I_n : \left| \frac{f(x_k + x_0) - f(x_0)}{x_k} - w \right| \ge \varepsilon \right\}$$

and so

$$\frac{1}{\mu_n} \left| \left\{ k \in J_n : \left| \frac{f(x_k + x_0) - f(x_0)}{x_k} - w \right| \ge \varepsilon \right\} \right|$$

$$\ge \frac{\lambda_n}{\mu_n} \frac{1}{\lambda_n} \left| \left\{ k \in I_n : \left| \frac{f(x_k + x_0) - f(x_0)}{x_k} - w \right| \ge \varepsilon \right\} \right|$$

for all $n \in \mathrm{IN}$. Now, using (2) and also taking the limit as $n \to \infty$ we get the proof.

(ii) Suppose the function $f: \mathrm{IR} \to \mathrm{IR}$ has λ—statistical derivative at a point. $x_0 \in \mathrm{IR}$ and let $\lim_{n\to\infty} \frac{\lambda_n}{\mu_n} = 1$. Since $I_n \subset J_n$ for any $\varepsilon > 0$, we may write,

$$\frac{1}{\mu_n} \left| \left\{ k \in J_n : \left| \frac{f(x_k+x_0)-f(x_0)}{x_k} - w \right| \ge \varepsilon \right\} \right|$$

$$= \frac{1}{\mu_n} \left| \left\{ n - \mu_n + 1 \le k \le n - \lambda_n : \left| \frac{f(x_k+x_0)-f(x_0)}{x_k} - w \right| \ge \varepsilon \right\} \right|$$

$$+ \frac{1}{\mu_n} \left| \left\{ k \in I_n : \left| \frac{f(x_k+x_0)-f(x_0)}{x_k} - w \right| \ge \varepsilon \right\} \right|$$

$$\le \frac{\mu_n - \lambda_n}{\mu_n} + \frac{1}{\mu_n} \left| \left\{ k \in I_n : \left| \frac{f(x_k+x_0)-f(x_0)}{x_k} - w \right| \ge \varepsilon \right\} \right|$$

$$\le \left(1 - \frac{\lambda_n}{\mu_n}\right) + \frac{1}{\lambda_n} \left| \left\{ k \in I_n : \left| \frac{f(x_k+x_0)-f(x_0)}{x_k} - w \right| \ge \varepsilon \right\} \right|$$

Since $\lim_{n\to\infty} \frac{\lambda_n}{\mu_n} = 1$, taking the limit as $n \to \infty$ we get the proof. tu.

Taking $\mu_n = n$ for every $n \in \mathrm{IN}$, from Theorem 3, we get the following results.

Corollary 1. Let a function $f: \mathrm{IR} \to \mathrm{IR}$ be given and $\lambda = (\lambda_n) \in \Lambda$. Then, we have the following.

(i) If a function $f: \mathrm{IR} \to \mathrm{IR}$ has a statistical derivative at a point $x_0 \in \mathrm{IR}$, then it has λ—statistical derivative at the same point in the case

$$\liminf_{n\to\infty} \frac{\lambda_n}{n} > 0 \tag{3}$$

holds.

(ii) If f has λ—statistical derivative then, f has μ—statistical derivative in the case

$$\lim_{n\to\infty} \frac{\lambda_n}{n} = 1 \tag{4}$$

holds.

Defnition 2 It is said that a function $f: \text{IR} \to \text{IR}$ has (V, λ)—derivative $w \in \text{IR}$ at a point $x_0 \in \text{IR}$ if

$$\lim_{n\to\infty} \frac{1}{\lambda_n} \sum_{k\in I_n} \frac{f(x_k + x_0) - f(x_0)}{x_k} = w$$

holds whenever $\chi_n = 0$ and $\lim_{n\to\infty} \chi_n = 0$.

A coequal de nition of this De nition is as follows.
A function $f: \text{IR} \to \text{IR}$ has a (V, λ)—derivative $w \in \text{IR}$ at a point $x_0 \in \text{IR}$ if

$$\lim_{n\to\infty} \frac{1}{\lambda_n} \sum_{k\in I_n} \frac{f(x_k + x_0) - f(x_0 - x_k)}{2x_k} = w$$

holds, whenever $\chi_n = 0$ and $\lim_{n\to\infty} \chi_n = 0$.

Remark 2. If we take $\lambda_n = n$, the above De nition coincides with the de nition of Ces ro derivative.

Defnition 3 It is said that a function $f: \text{IR} \to \text{IR}$ has strong (V, λ)—derivative $w \in \text{IR}$ at a point $x_0 \in \text{IR}$ if

$$\lim_{n\to\infty} \frac{1}{\lambda_n} \sum_{k\in I_n} \left| \frac{f(x_k + x_0) - f(x_0)}{x_k} - w \right| = 0$$

holds whenever $\chi_n = 0$ and $\lim_{n\to\infty} \chi_n = 0$.

A coequal de nition of this De nition is as follows.
A function $f: \text{IR} \to \text{IR}$ has strong (V, λ)—derivative $w \in \text{IR}$ at a point $x_0 \in \text{IR}$ if

$$\lim_{n\to\infty} \frac{1}{\lambda_n} \sum_{k\in I_n} \left| \frac{f(x_k + x_0) - f(x_0 - x_k)}{2x_k} - w \right| = 0$$

holds whenever $\chi_n = 0$ and $\lim_{n\to\infty} \chi_n = 0$.

Remark 3. If we take $\lambda_n = 0$, the overhead de nition coincides with the de nition of strong Ces ro derivative.

Theorem 4 If a function $f: \text{IR} \to \text{IR}$ has a strong (V, λ)—derivative at a point $x_0 \in \text{IR}$, then f has (V, λ)—derivative at x_0.

Proof. Suppose $f: \mathrm{IR} \to \mathrm{IR}$ has strong (V, λ)—derivative at a point $x_0 \in \mathrm{IR}$. Then, there exists some $w \in \mathrm{IR}$ such that

$$\lim_{n \to \infty} \frac{1}{\lambda_n} \sum_{k \in I_n} \left| \frac{f(x_k + x_0) - f(x_0)}{x_k} - w \right| = 0$$

Now, we may write

$$\frac{1}{\lambda_n} \sum_{k \in I_n} \frac{f(x_k + x_0) - f(x_0)}{x_k} - w = \frac{1}{\lambda_n} \sum_{k \in I_n} \left(\frac{f(x_k + x_0) - f(x_0)}{x_k} - w \right) \quad (5)$$

On the other hand, using generalized triangle inequaliy we also have

$$\frac{1}{\lambda_n} \left| \sum_{k \in I_n} \left(\frac{f(x_k + x_0) - f(x_0)}{x_k} - w \right) \right| \leq \frac{1}{\lambda_n} \sum_{k \in I_n} \left| \frac{f(x_k + x_0) - f(x_0)}{x_k} - w \right| \quad (6)$$

Since f has a strong (V, λ)—derivative at the point x_0, the right-hand side of the above inequality (6) tends to 0 and so that the left-hand side of this inequality tends to 0 as $n \to \infty$, that is

$$\lim_{n \to \infty} \frac{1}{\lambda_n} \sum_{k \in I_n} \left(\frac{f(x_k + x_0) - f(x_0)}{x_k} - w \right) = 0$$

From equality (5), we have

$$\lim_{n \to \infty} \frac{1}{\lambda_n} \sum_{k \in I_n} \frac{f(x_k + x_0) - f(x_0)}{x_k} = w$$

Hence, the function f has (V, λ)—derivative at x_0, which completes the proof. tu.

Defnition 4 A function $f: \mathrm{IR} \to \mathrm{IR}$ is (V, λ)—continuous at a point $x_0 \in \mathrm{IR}$ if (V, λ)—$\lim f(x_0 + x_n) = f(x_0)$ holds for each sequence $(x_n) \to 0$.

Theorem 5 If a function $f: \mathrm{IR} \to \mathrm{IR}$ is (V, λ)—di erentiable at a point x_0, then it is (V, λ)—continuous at x_0.

Proof. Let (x_n) be a sequence such that $\lim_{n \to \infty} \chi_n = 0$. Then, we must have. (V, λ)—$\lim x_n = 0$. Consequently, the following expression

$$f(x_0 + x_n) - f(x_0) = \frac{f(x_0 + x_n) - f(x_0)}{x_n} x_n$$

implies

$$(V, \lambda) - \lim(f(x_0 + x_n) - f(x_0)) = (V, \lambda) - \lim \frac{f(x_0 + x_n) - f(x_0)}{x_n} \cdot (V, \lambda) - \lim x_n = 0$$

Hence, (V, λ)—$\lim f(x_0 + x_n) = f(x_0)$, i.e. f is (V, λ)—continuous at x_0, and the proof is complete.tu.

Theorem 6 If a function $f: \mathrm{IR} \to \mathrm{IR}$ has a strong (V, λ)—derivative at a point $x_0 \in \mathrm{IR}$, then it has λ—statistical derivative at the point x_0.

Proof. Suppose $f: \mathrm{IR} \to \mathrm{IR}$ has strong (V, λ)—derivative at a point $x_0 \in \mathrm{IR}$. Then, for any $\varepsilon > 0$ there exists some $w \in \mathrm{IR}$ such that

$$\lim_{n \to \infty} \frac{1}{\lambda_n} \sum_{k \in I_n} \left| \frac{f(x_k + x_0) - f(x_0)}{x_k} - w \right| = 0 \qquad (7)$$

We define the sets $H_n = \left\{ k \in I_n : \left| \frac{f(x_k + x_0) - f(x_0)}{x_k} - w \right| \geq \varepsilon \right\}$ and $G_n = \left\{ k \in I_n \left| \frac{f(x_k + x_0) - f(x_0)}{x_k} - w \right| < \varepsilon \right\}$. Now, for any $\varepsilon > 0$

$$\frac{1}{\lambda_n} \sum_{k \in I_n} \left| \frac{f(x_k + x_0) - f(x_0)}{x_k} - w \right|$$

$$= \frac{1}{\lambda_n} \sum_{k \in H_n} \left| \frac{f(x_k + x_0) - f(x_0)}{x_k} - w \right|$$

$$+ \frac{1}{\lambda_n} \sum_{k \in G_n} \left| \frac{f(x_k + x_0) - f(x_0)}{x_k} - w \right|$$

$$\geq \frac{1}{\lambda_n} \sum_{k \in H_n} \left| \frac{f(x_k + x_0) - f(x_0)}{x_k} - w \right|$$

$$\geq \frac{1}{\lambda_n} \left| \left\{ k \in I_n : \left| \frac{f(x_k + x_0) - f(x_0)}{x_k} - w \right| \geq \varepsilon \right\} \right| \varepsilon \ominus$$

From (7) and the above inequation we have

$$\lim_{n \to \infty} \frac{1}{\lambda_n} \left| \left\{ k \in I_n : \left| \frac{f(x_k + x_0) - f(x_0)}{x_k} - w \right| \geq \varepsilon \right\} \right| = 0 \cdot$$

Hence, f has λ—statistical derivative at x_0.tu.

If we take $\lambda_n = 0$ for every $n \in \mathrm{IN}$, then from Theorem 6, we get the following result which is Theorem 3.1(i) in [6].

Corollary 2. If a function $f: \mathrm{IR} \to \mathrm{IR}$ has a strong Cesàro derivative at a point $x_0 \in \mathrm{IR}$, then it has a statistical derivative at the point x_0.

Theorem 7 If a function $f: \mathrm{IR} \to \mathrm{IR}$ has λ—statistical derivative at a point $x_0 \in \mathrm{IR}$, then it has strong (V, λ)—derivative at x_0 provided that $((f(x_k + x_0) - f(x_0))/x_k)$ is bounded for any sequence (x_n) such that $x_n > 0$ for every n and $\lim_n \to \infty\, x_n = 0$.

Proof. Suppose $f: \mathrm{IR} \to \mathrm{IR}$ has λ—statistical derivative at a point $x_0 \in \mathrm{IR}$. Then, for any $\varepsilon > 0$ there exists some $w \in \mathrm{IR}$ such that

$$\lim_{n \to \infty} \frac{1}{\lambda_n} \left| \left\{ k \in I_n : \left| \frac{f(x_k + x_0) - f(x_0)}{x_k} - w \right| \geq \varepsilon \right\} \right| = 0 \tag{8}$$

Now, since $\left(\frac{f(x_k + x_0) - f(x_0)}{x_k} \right)$ is bounded, there exists a number $A > 0$ such that $\left| \frac{f(x_k + x_0) - f(x_0)}{x_k} - w \right| \leq A$ for every $k \in \mathrm{IN}$. Now, for any $\varepsilon > 0$

$$\frac{1}{\lambda_n} \sum_{k \in I_n} \left| \frac{f(x_k + x_0) - f(x_0)}{x_k} - w \right|$$

$$= \frac{1}{\lambda_n} \sum_{k \in H_n} \left| \frac{f(x_k + x_0) - f(x_0)}{x_k} - w \right|$$

$$+ \frac{1}{\lambda_n} \sum_{k \in G_n} \left| \frac{f(x_k + x_0) - f(x_0)}{x_k} - w \right|$$

$$\leq \frac{A}{\lambda_n} \sum_{k \in H_n} 1 + \frac{1}{\lambda_n} \sum_{k \in G_n} \left| \frac{f(x_k + x_0) - f(x_0)}{x_k} - w \right|$$

$$\leq \frac{A}{\lambda_n} \left| \left\{ k \in I_n : \left| \frac{f(x_k + x_0) - f(x_0)}{x_k} - w \right| \geq \varepsilon \right\} \right|$$

$$+ \frac{1}{\lambda_n} \sum_{k \in I_n} \varepsilon$$

From (8) and the above inequality, we have

$$\lim_{n \to \infty} \frac{1}{\lambda_n} \sum_{k \in I_n} \left| \frac{f(x_k + x_0) - f(x_0)}{x_k} - w \right| = 0$$

Hence, f has a strong (V, λ)—derivative at x_0.tu.

If we take $\lambda_n = 0$ for every $n \in \mathrm{IN}$, then from Theorem 7, we get the following result, which is Theorem 3.1 (ii) in [6].

Corollary 3. If a function $f: \mathrm{IR} \to \mathrm{IR}$ has a statistical derivative at a point $x_0 \in \mathrm{IR}$, then it has strong Ces ro derivative at x_0 provided that $\left(\frac{f(x_k + x_0) - f(x_0)}{x_k} \right)$ is bounded.

Theorem 8 Let $\lambda = (\lambda_n)$ and $\mu = (\mu_n)$ be two sequences in Λ such that $\lambda_n \leq \mu_n$ for all $n \in \mathrm{IN}$.

(i) If a function $f: \mathrm{IR} \to \mathrm{IR}$ has strong (V, μ)—derivative, then f has strong (V, λ)—derivative in the case

$$\liminf_{n \to \infty} \frac{\lambda_n}{\mu_n} > 0 \tag{9}$$

(ii) If a function $f: \mathrm{IR} \to \mathrm{IR}$ has strong (V, λ)—derivative, then f has strong (V, μ)—derivative in the case

$$\lim_{n \to \infty} \frac{\lambda_n}{\mu_n} = 1 \tag{10}$$

holds.

Proof. (i) Suppose that $f: \mathrm{IR} \to \mathrm{IR}$ has a strong (V, μ)—derivative at a point $x_0 \in \mathrm{IR}$ and let $\lambda_n \leq \mu_n$ for all $n \in \mathrm{IN}$, so $I_n \subset J_n$, where $I_n = [n-\lambda_n + 1, n]$ and $J_n = [n-\mu_n + 1, n]$. Then, there exists some $w \in \mathrm{IR}$ such that

$$\lim_{n \to \infty} \frac{1}{\mu_n} \sum_{k \in J_n} \left| \frac{f(x_k + x_0) - f(x_0)}{x_k} - w \right| = 0 \tag{11}$$

Now, since $I_n \subset J_n$ we can write

$$\sum_{k \in J_n} \left| \frac{f(x_k + x_0) - f(x_0)}{x_k} - w \right| \geq \sum_{k \in I_n} \left| \frac{f(x_k + x_0) - f(x_0)}{x_k} - w \right|$$

and so that

$$\frac{1}{\mu_n} \sum_{k \in J_n} \left| \frac{f(x_k + x_0) - f(x_0)}{x_k} - w \right| \geq \frac{\lambda_n}{\mu_n} \frac{1}{\lambda_n} \sum_{k \in I_n} \left| \frac{f(x_k + x_0) - f(x_0)}{x_k} - w \right| \tag{12}$$

for all $n \in \mathrm{IN}$. Using (9) and (11) and also taking the limit as $n \to \infty$ in (12), we get that f has a strong (V, λ)—derivative.

(ii) Suppose the function $f: \mathrm{IR} \to \mathrm{IR}$ has a strong (V, λ)—derivative at a point $x_0 \in \mathrm{IR}$ and let $\lim_{n \to \infty} \frac{\lambda_n}{\mu_n} = 1$. Then, there exists some $w \in \mathrm{IR}$ such that

$$\lim_{n \to \infty} \frac{1}{\lambda_n} \sum_{k \in I_n} \left| \frac{f(x_k + x_0) - f(x_0)}{x_k} - w \right| = 0 \tag{13}$$

Since $I_n \subset J_n$ for every n, we may write,

$$\frac{1}{\mu_n} \sum_{k \in J_n} \left| \frac{f(x_k + x_0) - f(x_0)}{x_k} - w \right|$$

$$= \frac{1}{\mu_n} \sum_{k=n-\mu_n+1}^{n-\lambda_n} \left| \frac{f(x_k + x_0) - f(x_0)}{x_k} - w \right|$$

$$+ \frac{1}{\mu_n} \sum_{k \in I_n} \left| \frac{f(x_k + x_0) - f(x_0)}{x_k} - w \right|$$

$$\leq \frac{\mu_n - \lambda_n}{\mu_n} + \frac{1}{\mu_n} \sum_{k \in I_n} \left| \frac{f(x_k + x_0) - f(x_0)}{x_k} - w \right|$$

$$\leq \left(1 - \frac{\lambda_n}{\mu_n} \right) + \frac{1}{\lambda_n} \sum_{k \in I_n} \left| \frac{f(x_k + x_0) - f(x_0)}{x_k} - w \right|$$

We get the proof by using (10) and (13) and taking the limit as $n \to \infty$ in this last inequality.tu.

3 Conclusion

The expansions of the recently introduced statistical derivative and Ces ro derivative to λ—statistical derivative, (V, λ)—statistical derivative, and strongly (V, λ)—statistical derivative are substantial steps in the development of the statistical di erentiation theory. These pristine derivatives can be applied to a broader range of problems in statistics and machine learning. Furthermore, this investigation will be a helpful citation for the examinations to be carried out in connected topics in the following stages and for the scientists studying in connected elds.

References

1. Fast H (1951) Sur la convergence statistique. Colloq Math 2:241–244
2. Connor JS (1988) The statistical and strong p-Ces ro convergence of sequences. Analysis 8:47–63
3. Çolak R (2011) On λ—statistical convergence. In: Conference on summability and applications, Istanbul Commerce University, May 12–13, stanbul, T rkiye
4. Schoenberg IJ (1959) The integrability of certain functions and related summabilitymethods. Amer Math Monthly 66, 361 375
5. Mursaleen M (2000) λ—statistical convergence. Math Slovaca 50:111–115
6. Nuray F (2020) Ces ro and statistical derivative. Facta Univ Ser Math 35(5):1393–1398

Overlapping Iterative Numerical Method for Solving System of Singularly Perturbed Convection-Diffusion Problems with Mixed-Type Boundary Conditions

J. Jenifa and J. Christy Roja

Abstract In this article, using an overlapping iterative numerical technique on a Shiskin-type mesh, we study the convergence of a system of singularly perturbed convection-diffusion equations (SSPCDEs) with mixed-type boundary conditions (MBCs). The study is based on the definition of a few auxiliary problems that demonstrate the method's uniform convergence in two steps and distinguish between the discretization and iteration errors. Using the supremum norm, an error estimate of order $O(N^{-1}\ln^2 N)$ is deduced. The hypothetical results are demonstrated by numerical experiments.

Keywords Singularly perturbed problems · Convection-diffusion equations · Iterative method · Hybrid difference scheme · Mixed-type boundary conditions

1 Introduction

Inspired by [1, 2], we examine the (SPPs) of determining $u_1, u_2 \in C^1(\bar{\Omega}) \cap C^2(\Omega)$ such that

$$
\begin{cases}
\mathcal{L}_1\mathbf{u} \equiv -\epsilon u_1'' + a_1(t)u_1' + b_{11}(t)u_1 + b_{12}(t)u_2 = f_1(t), \\
\mathcal{L}_2\mathbf{u} \equiv -\epsilon u_2'' + a_2(t)u_2' + b_{21}(t)u_1 + b_{22}(t)u_2 = f_2(t), \quad t \in \Omega = (0, 1) \\
\mathcal{B}_{10}u_1(0) \equiv \beta_{10}u_1(0) - \beta_{11}u_1'(0) = \mathcal{A}_1, \\
\mathcal{B}_{20}u_2(0) \equiv \beta_{20}u_2(0) - \beta_{21}u_2'(0) = \mathcal{A}_2, \\
\mathcal{B}_{11}u_1(1) \equiv \gamma_{11}u_1(1) + \epsilon\gamma_{12}u_1'(1) = \mathcal{B}_1, \\
\mathcal{B}_{21}u_2(1) \equiv \gamma_{21}u_2(1) + \epsilon\gamma_{22}u_2'(1) = \mathcal{B}_2,
\end{cases}
\tag{1}
$$

J. Jenifa · J. Christy Roja (✉)
Department of Mathematics, St. Joseph's College, Affiliated to Bharathidasan University, Thiruchirappalli 620002, Tamil Nadu, India
e-mail: jchristyrojaa@gmail.com

© The Author(s), under exclusive license to Springer Nature Singapore Pte Ltd. 2024
D. Giri et al. (eds.), *Proceedings of the Tenth International Conference on Mathematics and Computing*, Lecture Notes in Networks and Systems 964,
https://doi.org/10.1007/978-981-97-2066-8_18

where $0 < \epsilon \ll 1$. Effectively smooth functions on $\bar{\Omega}$ include $a_i(t), b_{ij}(t),$ and $f_i(t), \ i, j = 1, 2$.

The given coupled system's matrix-vector representation is written as

$$L\mathbf{u} \equiv \begin{pmatrix} \mathcal{L}_1 u \\ \mathcal{L}_2 u \end{pmatrix} \equiv -E u'' + A(t)u' + B(t)u = \mathbf{f}(t), \ t \in \Omega,$$

under boundary conditions

$$\mathbf{B}_0\mathbf{u}(0) = \begin{pmatrix} \mathcal{B}_{10}u_1(0) \\ \mathcal{B}_{20}u_2(0) \end{pmatrix} = T_1\mathbf{u}(0) - T_2\mathbf{u}'(0),$$

$$\mathbf{B}_1\mathbf{u}(1) = \begin{pmatrix} \mathcal{B}_{11}u_1(1) \\ \mathcal{B}_{21}u_2(1) \end{pmatrix} = T_3\mathbf{u}(1) + \mathcal{E}T_4\mathbf{u}'(1),$$

where $E = \text{diag}(\epsilon, \epsilon), \quad T_1 = \text{diag}(\beta_{10}, \beta_{20}), \ T_2 = \text{diag}(\beta_{11}, \beta_{21}), \ T_3 = \text{diag}(\gamma_{11}, \gamma_{21}), \quad T_4 = \text{diag}(\gamma_{12}, \gamma_{22}), \ \mathbf{f}(x) = (f_1, f_2)^T,$ and" $\mathbf{u} = (u_1, u_2)^T$. Matrix $A(t) = \text{diag}(a_1(t), a_2(t))$ It is assumed that the coupling matrix $B(t) = (b_{ij}(t))_{2 \times 2}$ satisfied the subsequent requirements:

$$\begin{cases} a_i(t) \geq \alpha_i \geq 1, \ b_{12}(t) \leq 0, \ b_{21}(t) \leq 0 \\ \text{and } \sum_{j=1}^{2} b_{ij}(t) \geq \beta > 1, \ i = 1, 2, \ \forall t \in \Omega. \text{ Let } \alpha = \min(\alpha_1, \alpha_2) \text{ and } \alpha \geq 1. \\ \tau_1 = \min(\beta_{10}, \beta_{20}), \ \tau_2 = \min(\beta_{11}, \beta_{21}), \ \tau_3 = \min(\gamma_{11}, \gamma_{21}) \text{ and } \tau_4 = \min(\gamma_{12}, \gamma_{22}). \\ \gamma_{j1} + \gamma_{j2} > \lambda_j, \ j = 1, 2 \text{ where } \lambda_j > 0 \text{ and } \lambda' = \min\{\lambda_1, \lambda_2\}. \end{cases} \quad (2)$$

Also $\beta_{j0}, \beta_{j1} > 0, \quad \beta_{j0} - \beta_{j1} > \mu_j, \quad j = 1, 2$ where $\mu_j > 0$ and $\mu' = \min\{\mu_1, \mu_2\}$, and $2\gamma_{j1} + \epsilon\gamma_{j2} > 0, \ j = 1, 2$.

Problem (1) appears to have a distinct solution with an overlapping boundary layer on the right side under these assumptions. These kinds of issues occur in some resistor-capacitor devices and in optimum control issues (see [3]). It is widely recognized that SPPs can be solved using a wide variety of numerical techniques (see, for example [4, 5]).

The SSPBVPs have attracted the attention of many investigators during the last few decades. In [6–8], the special mesh technique was applied to solve SPPs. The adaptive grid approach in [9] has been evaluated by the authors, who have also confirmed its feasibility and efficacy when applied to both linear and non-linear numerical problems.

In [10] discussed uniformly convergent numerical methods under mixed-type boundary condition SSPCDP. A parameter uniform numerical technique for SSPCDEs with MBCs on smooth data has been established by Tamilselvan et al. [11].

The development of robust numerical techniques for SPRCDPs has received a lot of attention since 2003. Applications for this set of equations include predator-prey population flow and electroanalytic chemistry. At large Reynolds numbers, conversely, a reaction-diffusion system arises through linearized Navier-Stokes equations. However, since non-iterative techniques are the focus of the majority of methods created for SSPPs, it makes sense that the authors of this study would want to design an iterative method approach for a problem of that nature.

The remaining portions of this article have been structured as follows: A few derivatives of the analytical solution to the mixed-type boundary value issue with singular perturbations are shown in Sect. 2. Section 3 explains the numerical strategy. A more detailed analysis of the errors associated with the method is presented in Sect. 4. Section 5 provides numerical examples, while Sect. 6 offers conclusions.

Notations: We refer to C throughout the study as a general positive constant that is unaffected by the discretization parameter N, the perturbation parameter ϵ, and the iteration parameter k.

Let $\mathbf{u} : D \to R$, $D \subseteq R$. When examining how a numerical solution for a singularly perturbed issue approaches the precise answer, the proper norm to use is $||u||_D = \sup_{t \in D} |u(t)|$. $\mathfrak{z} = (z_1, z_2)^T$ is a vector-valued function and define $||\mathfrak{z}||_\Omega = \max\{||z_1||_\Omega, ||z_2||_\Omega\}$. Regarding a mesh function vector $Z(t_i) = (Z_1(t_i), Z_2(t_i))^T$, define $||Z||_{\Omega^N} = \max_{q=1,2}\left(\max_{t_i \in \Omega^N} |Z_q(t_i)|\right)$. Additionally specify $||\mathbf{g}||_\infty = \max_{q=1,2} |g_q|$.

2 Analytical Results

We illustrated the derivatives of the continuous problem's analytical solution in this section because they are crucial to the examination of the discrete problem shown in Sect. 3.

In order to show that breaking down the solution $\mathbf{u}$ into regular and singular parts yields the sharper bounds on the derivatives of the solutions, $\mathbf{u} = \mathbf{v} + \mathbf{w}$ where $\mathbf{v} = (v_1, v_2)^T$ and $\mathbf{w} = (w_1, w_2)^T$. The form of the regular part $\mathbf{v}$ is as follows: $\mathbf{v} = v_0 + \epsilon v_1$ where $v_0 = (v_{01}, v_{02})^T$ and $v_1 = (v_{11}, v_{12})^T$ are defined as the solutions to the problems, respectively.

$$A(t)v_0' + B(t)v_0 = \mathbf{f}(t), \quad t \in \Omega,$$

$$\begin{pmatrix} \mathcal{B}_{10}v_{01}(0) \\ \mathcal{B}_{20}v_{02}(0) \end{pmatrix} = \begin{pmatrix} \mathcal{B}_{10}u_1(0) \\ \mathcal{B}_{20}u_2(0) \end{pmatrix},$$

$$\text{and } A(t)v_1' + B(t)v_1 = \begin{pmatrix} \frac{d^2}{dt^2} & 0 \\ 0 & \frac{d^2}{dt^2} \end{pmatrix} v_0, \quad t \in \Omega,$$

$$\begin{pmatrix} \mathcal{B}_{10}v_{11}(0) \\ \mathcal{B}_{20}v_{12}(0) \end{pmatrix} = \mathbf{0}.$$

As a result, the non-homogeneous problem $L\mathbf{v} = \mathbf{f}$ has a regular part $\mathbf{v}$.

Following that, the homogeneous problem's solution is represented by the singular part $\mathbf{w}$,

$$L\mathbf{w} = \mathbf{0}, \quad \begin{pmatrix} B_{10}w_1(0) \\ B_{20}w_2(0) \end{pmatrix} = \mathbf{0} \text{ and } \begin{pmatrix} B_{11}w_1(1) \\ B_{21}w_2(1) \end{pmatrix} = \begin{pmatrix} B_{11}u_1(1) - B_{11}v_{01}(1) \\ B_{21}u_2(1) - B_{21}v_{02}(1) \end{pmatrix}.$$

The bounds on the derivatives of the regular part and singular part of the solution $\mathbf{u}$ of (1) is provided by the lemma that follows.

Lemma 1 *For $0 \le s < 4$, the derivatives of the solution $\mathbf{u}$ of (1) attain the following bounds where $\mathbf{u} = \mathbf{v} + \mathbf{w}$,*

$$||\mathbf{v}^{(s)}||_{\bar{\Omega}} \le C(1 + \epsilon^{2-s}) \quad and \quad \left| w_n^{(s)}(t) \right| \le C\epsilon^{-s}e^{-\alpha(1-t)/\epsilon}, \quad \forall t \in \bar{\Omega}, \quad n = 1, 2.$$

Proof Using the proof methodology from [12], it is possible to prove the display lemma.   □

3 Discrete Problem

Two overlapping subdomains of the domain $\Omega = (0, 1)$ are identified as follows: $\Omega_l = (0, 1 - \sigma)$ and $\Omega_r = (1 - 2\sigma, 1)$, where the subdomain's transition value σ is taken as

$$\sigma = \min\left\{ \frac{1}{3}, \frac{2\epsilon}{\alpha} \ln N \right\} \tag{3}$$

as in Ref. [13] (Fig. 1).

Let $N = 2^N$, $n \ge 2$. In every subdomain $\Omega_p = (a, d)$, where $p = \{l, r\}$, establish a uniform mesh(or grid) length $h_p = (d - a)/N$. In $\bar{\Omega}_p^N$, let $a = t_0 < t_1 < \ldots < t_{N-1} < t_N = d$ with $t_j = a + jh_p$, $j = 0, \ldots, N$.

We employ the midpoint difference scheme with a uniform mesh on Ω_l and the central finite difference scheme on Ω_r in the suggested system. Next, the equivalent discretization inside every subdomain Ω_p^N, $p = \{l, r\}$, has the following form:

$$L^N\mathbf{U}_l(t_i) = \begin{cases} \mathcal{L}_1^N\mathbf{U}_l(t_i) = -\epsilon\delta^2 U_1(t_i) + a_{1,i-1/2}D^-U_1(t_i) \\ \quad + b_{11,i-1/2}\hat{U}_1(t_i) + b_{12,i-1/2}\hat{U}_2(t_i) = f_{1,i-1/2}, \\ \mathcal{L}_2^N\mathbf{U}_l(t_i) = -\epsilon\delta^2 U_2(t_i) + a_{2,i-1/2}D^-U_2(t_i) \\ \quad + b_{21,i-1/2}\hat{U}_1(t_i) + b_{22,i-1/2}\hat{U}_2(t_i) = f_{2,i-1/2}, i = 1, \ldots, N-1 \end{cases} \tag{4}$$

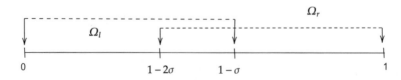

Fig. 1 Breakdown of the initial computing domain

and

$$
L^N \mathbf{U}_r(t_i) =
\begin{cases}
\mathcal{L}_1^N U_r(t_i) = -\epsilon \delta^2 U_1(t_i) + a_{1,i} D^0 U_1(t_i) + b_{11,i} U_1(t_i) + b_{12,i} U_2(t_i) = f_{1,i} \\
\mathcal{L}_2^N U_r(t_i) = -\epsilon \delta^2 U_2(t_i) + a_{2,i} D^0 U_2(t_i) + b_{21,i} U_1(t_i) \\
\quad + b_{22,i} U_2(t_i) = f_{2,i}, \ i = 1, \ldots, N-1
\end{cases}
\tag{5}
$$

where $\delta^2 U_j(t_i) = \dfrac{1}{h_p^2} \left(U_j(t_{i-1}) - 2U_j(t_i) + U_j(t_{i+1}) \right)$,

$D^- U_j(t_i) = \dfrac{U_j(t_i) - U_j(t_{i-1})}{h_l}$, $\hat{U}_j(t_i) \equiv \dfrac{(U_j(t_i) + U_j(t_{i-1}))}{2}$,

$D^0 U_j(t_i) = \dfrac{U_j(t_{i+1}) - U_j(t_{i-1})}{2h_r}$, where $a_{j,i-1/2} \equiv \dfrac{a(t_{i-1} + t_i)}{2}$, $a_{j,i} \equiv a(t_i)$;

similarly for $b_{1j,i-1/2}$, $b_{2j,i-1/2}$, $f_{j,i-1/2}$, $b_{1j,i}$, $b_{2j,i}$ and $f_{j,i}$, $j = 1, 2$.
The discrete problem is $L^N \mathbf{U}_p(t_i) = \mathsf{F}(t_i)$, where

$$
\mathsf{F}(t_i) =
\begin{cases}
\mathbf{f}_{i-\frac{1}{2}}, & t_i \in \bar{\Omega}_l^N, \\
\mathbf{f}_i, & t_i \in \bar{\Omega}_r^N.
\end{cases}
$$

For the discrete problem (4)–(5), we present a discrete overlapping iterative numerical technique in four steps.

step 1. Select the first mesh function.
$\mathbf{U}^{[0]}(t_i) \equiv \mathbf{0}$, $0 < t_i < 1$, $\mathbf{B}_0 \mathbf{U}^{[0]}(0) = \mathbf{B}_0 \mathbf{u}(0)$, $\mathbf{B}_1 \mathbf{U}^{[0]}(1) = \mathbf{B}_1 \mathbf{u}(1)$.

step 2. The following discrete problems have mesh functions $\mathbf{U}_p^{[k]}$, $p = \{l, r\}$ that we compute.

$$
\begin{aligned}
L^N \mathbf{U}_r^{[k]} &= \mathbf{f}_i \quad \text{in } \Omega_r^N, \ \mathbf{U}_r^{[k]}(1 - 2\sigma) = I \mathbf{U}_l^{[k-1]}(1 - 2\sigma), \ \mathbf{B}_1 \mathbf{U}_r^{[k]}(1) = \mathbf{B}_1 \mathbf{u}(1), \\
L^N \mathbf{U}_l^{[k]} &= \mathbf{f}_{i-\frac{1}{2}} \quad \text{in } \Omega_l^N, \ \mathbf{B}_0 \mathbf{U}_l^{[k]}(0) = \mathbf{B}_0 \mathbf{u}(0), \ \mathbf{U}_l^{[k]}(1 - \sigma) = I \mathbf{U}_r^{[k]}(1 - \sigma)
\end{aligned}
$$

where $I \mathbf{U}_p^{[k]}$, $k \geq 1$ represents the linear interpolant piecewise of $\mathbf{U}_p^{[k]}$ on the mesh $\bar{\Omega}_p^N = (\bar{\Omega}_r^N \setminus \bar{\Omega}_l) \cup \bar{\Omega}_l^N$.

step 3. Combining the solutions on the subdomains, we estimate the mesh function $\mathbf{U}^{[k]}$.

$$
\mathbf{U}^{[k]}(t_i) =
\begin{cases}
\mathbf{U}_r^{[k]}(t_i), & \text{if } t_i \in \bar{\Omega}_r^N \setminus \bar{\Omega}_l, \\
\mathbf{U}_l^{[k]}(t_i), & \text{if } t_i \in \bar{\Omega}_l^N.
\end{cases}
\tag{6}
$$

step 4. Proceed to Step 2 if the halting requirement $\|\mathbf{U}^{[k+1]} - \mathbf{U}^{[k]}\|_{\bar{\Omega}^N} \leq \mathbf{tol}$ is not met. In this case, the user-recommended accuracy is **tol**.

Since the matrix associated with L^N for each Ω_p, $p = \{l, r\}$ is an M-matrix, it meets the next discrete maximum principle.

Lemma 2 (*Discrete maximum principle*) *For any mesh function* $\mathbf{U}_p(t_i)$ *suppose that* $B_0^N \mathbf{U}_p(t_0) \geq 0$, $B_1^N \mathbf{U}_p(t_N) \geq 0$, *and* $L^N \mathbf{U}_p(t_i) \geq \mathbf{0}$. *Then* $\mathbf{U}_p(t_i) \geq \mathbf{0}$, $\forall\ i = 0, 1, \cdots, N$.

This lemma has the following stability conclusion as an immediate effect.

Lemma 3 *If* $U_j(t_i)$ *is any mesh function then for all* $t_i \in \bar{\Omega}_p^N$,

$$|U_j(t_i)| \leq C \max\left\{|B_0^N U_1(t_0)|, |B_1^N U_1(t_N)|, |B_0^N U_2(t_0)|, |B_1^N U_2(t_N)|, \|\mathcal{L}_1^N \mathbf{U}\|_{\Omega_p^N}, \|\mathcal{L}_2^N \mathbf{U}\|_{\Omega_p^N}\right\}, \text{ where } j = 1, 2.$$

Proof Please refer to [14].   □

4   Error Estimate

This part presents an error estimate of the numerical approach that is overlapping and iterative, as it is explained in Sect. 3. The estimate is provided with the intention of identifying a few auxiliary issues. It aids the writers in convincingly demonstrating the method's uniform convergence in two stages by separating the errors associated with discretization and iteration.

We define the auxiliary problems as follows: In order to obtain the mesh function $\widetilde{\mathbf{U}}_p$, $p = \{l, r\}$

$$L^N \widetilde{\mathbf{U}}_r = \mathbf{f}_i \text{ in } \Omega_r^N, \quad \widetilde{\mathbf{U}}_r(1 - 2\sigma) = \mathbf{u}(1 - 2\sigma), \quad \mathbf{B}_1 \widetilde{\mathbf{U}}_r(1) = \mathbf{B}_1 \mathbf{u}(1),$$
$$L^N \widetilde{\mathbf{U}}_l = \mathbf{f}_{i-\frac{1}{2}} \text{ in } \Omega_l^N, \quad \mathbf{B}_0 \widetilde{\mathbf{U}}_l(0) = \mathbf{B}_0 \mathbf{u}(0), \quad \widetilde{\mathbf{U}}_l(1 - \sigma) = \mathbf{u}(1 - \sigma).$$

For $L^N \widetilde{\mathbf{U}}_p$, $p = \{l, r\}$, the boundary conditions make use of the exact solution $\mathbf{u}$ of (1).

Lemma 4 *Suppose* $\mathbf{u}$ *is the solution of (1) and* $\widetilde{\mathbf{U}}_p$, $p = \{l, r\}$ *are solutions of the above-mentioned auxiliary problems. Then* $\|\widetilde{\mathbf{U}}_p - \mathbf{u}\|_{\bar{\Omega}_p^N} \leq C N^{-1} \ln^2 N$.

Proof Case (i): Estimating error bounds on $\bar{\Omega}_r^N$.
From [14], we can derive the truncation error estimate on $t_i \in \bar{\Omega}_r^N$ as

$$\|(L^N - L)\mathbf{u}\|_{\Omega_r^N} \leq C \left(\epsilon h_r \|\mathbf{u}^{(3)}\|_{\Omega_r} + \|a_1\| h_r \|\mathbf{u}^{(2)}\|_{\Omega_r}\right). \tag{7}$$

We must decompose $\mathbf{u}$ as in Lemma 1 in order to get a bound on $\|L^N(\widetilde{\mathbf{U}}_r - \mathbf{u})\|_{\bar{\Omega}_r^N}$.

Consider that $\quad \|L^N(\widetilde{\mathbf{U}}_r - \mathbf{u})\|_{\Omega_r^N} = \|\mathbf{f} - L^N \mathbf{u}\|_{\Omega_r^N}$
$$\leq \|(L^N - L)\mathbf{v}\|_{\Omega_r^N} + \|(L^N - L)\mathbf{w}\|_{\Omega_r^N}. \tag{8}$$

Regarding the initial term situated on the right side of (8), using $\epsilon \leq CN^{-1}$, $h_r \leq CN^{-1}$ also Lemma 1, one can deduce that

$$\|(L^N - L)\mathbf{v}\|_{\Omega_r^N} \leq C \left(\epsilon h_r \|\mathbf{v}^{(3)}\|_{\Omega_r} + \|a_1\| h_r \|\mathbf{v}^{(2)}\|_{\Omega_r} \right)$$
$$\leq CN^{-1}.$$

Lemma 1 and $h_r \leq C\epsilon N^{-1} \ln N$, $\epsilon \leq CN^{-1}$ are used to find the 2nd term on the right side of (8),

$$\|(L^N - L)\mathbf{w}\|_{\Omega_r^N} \leq C \left(\epsilon h_r \|\mathbf{w}^{(3)}\|_{\Omega_r} + \|a_1\| h_r \|\mathbf{w}^{(2)}\|_{\Omega_r} \right)$$
$$\leq CN^{-1}\epsilon^{-1} \ln N.$$

With the estimations from (8) above, one can get

$$\|L^N(\tilde{\mathbf{U}}_r - \mathbf{u})\|_{\Omega_r^N} \leq CN^{-1} + C\epsilon^{-1}N^{-1} \ln N, \quad \text{for some } C$$

along with

$$|(\tilde{\mathbf{U}}_r - \mathbf{u}(1 - 2\sigma)| = 0 \quad \text{and} \quad |\mathbf{B}_1(\tilde{\mathbf{U}}_r - \mathbf{u})(1)| = 0.$$

Using appropriate barrier function and applying the discrete maximum principle for the operator L^N on $\bar{\Omega}_r^N$, one can derive that when $\sigma = \dfrac{2\epsilon}{\alpha} \ln N$

$$\|(\tilde{\mathbf{U}}_r - \mathbf{u})\|_{\bar{\Omega}_r^N} \leq CN^{-1} + C(x_i - (1 - 2\sigma))\epsilon^{-1}N^{-1} \ln N$$
$$\leq CN^{-1} \ln^2 N.$$

$$\text{Consequently, } \|(\tilde{\mathbf{U}}_r - \mathbf{u})\|_{\bar{\Omega}_r^N \setminus \bar{\Omega}_l} \leq CN^{-1} \ln^2 N. \tag{9}$$

Case (ii): Estimating error bounds on $\bar{\Omega}_l^N$.

At every point $t_i \in \bar{\Omega}_l^N$, we apply the solution decomposition as in Lemma 1, and the difference $(\tilde{\mathbf{U}}_l - \mathbf{u})$ can be defined in a way described below:

$$(\tilde{\mathbf{U}}_l - \mathbf{u})(t_i) = (\tilde{\mathbf{V}}_l - \mathbf{v})(t_i) + (\tilde{\mathbf{W}} - \mathbf{w}_l)(t_i). \tag{10}$$

We employ the midpoint difference technique on $\bar{\Omega}_l^N$ in this proposed scheme. Based on $t_i \in \bar{\Omega}_l^N$, the truncation error estimate found in [14] can be obtained as

$$\|(L^N - L)\mathbf{u}\|_{\Omega_l^N} \leq C\epsilon h_l \|\mathbf{u}^{(3)}\|_{\Omega_l} + Ch_l \left(\|\mathbf{u}^{(2)}\|_{\Omega_l} + \|\mathbf{u}\|_{\Omega_l} \right). \tag{11}$$

*Subcase (i):*Given the local truncation error estimation for the first phrase on the right (10) and Lemma 1, we obtain $\epsilon \leq CN^{-1}$, $h_l \leq CN^{-1}$.

$$\begin{aligned}
\|L^N(\tilde{\mathbf{V}}_l - \mathbf{v})\|_{\Omega_l^N} &= \|(L^N - L)\mathbf{v}\|_{\Omega_l^N} \\
&\leq C\epsilon h_l \|\mathbf{v}^3\|_{\Omega_l} + Ch_l\left(\|\mathbf{v}^2\|_{\Omega_l} + \|\mathbf{v}\|_{\Omega_l}\right) \\
&\leq CN^{-1}
\end{aligned}$$

along with $|\mathbf{B}_0(\tilde{\mathbf{V}}_l - \mathbf{v})(0)| = 0$ and $|(\tilde{\mathbf{V}}_l - \mathbf{v})(1 - \sigma)| = 0$.
With an appropriate barrier function and the discrete maximum principle applied to the operator L^N on $\bar{\Omega}_l^N$, it can be deduced that

$$\|\tilde{\mathbf{V}}_l - \mathbf{v}\|_{\bar{\Omega}_l^N} \leq CN^{-1}.$$

Subcase (ii): The second term on the right side of (10), when $\sigma = \dfrac{2\epsilon}{\alpha} \ln N$, utilizing the contentions examined as in [[15], Lemma(6)], one can get

$$\|\mathbf{W}_l - \mathbf{w}\|_{\bar{\Omega}_l^N} \leq CN^{-1}.$$

Right now, using the layer and regular parts error bounds, we obtain

$$\|\tilde{\mathbf{U}}_l - \mathbf{u}\|_{\bar{\Omega}_l^N} \leq CN^{-1}\ln^2 N. \tag{12}$$

We combine (9) with (12)to get the required approximation.

$$\|\tilde{\mathbf{U}} - \mathbf{u}\|_{\bar{\Omega}^N} \leq CN^{-1}\ln^2 N.$$

$\square$

We now provide some documentation that will be helpful in illustrating Lemma 5 intermediate result, which will be used to support Theorem 6 primary result.

$$\eta_{1-\sigma} = \|(\tilde{\mathbf{U}}_l - l\tilde{\mathbf{U}}_r)(1 - \sigma)\|_\infty$$
$$\eta_{1-2\sigma} = \|(\tilde{\mathbf{U}}_r - l\tilde{\mathbf{U}}_l)(1 - 2\sigma)\|_\infty.$$

Lemma 5 *Suppose* $\psi = \eta_{1-\sigma} + \eta_{1-2\sigma}$. *Then* $\psi \leq CN^{-1}\ln^2 N$.

Proof As $1 - \sigma \in \bar{\Omega}_l^N$ from Lemma 4, we have $\eta_{1-\sigma} \leq CN^{-1}\ln^2 N$.
We now estimate $\eta_{1-2\sigma}$. Using an inequality triangle, we obtain

$$\left|\left(\mathbf{u} - l\tilde{\mathbf{U}}_l\right)(1 - 2\sigma)\right| \leq \left|(\mathbf{u} - l\mathbf{u})(1 - 2\sigma)\right| + \left|l\left(\mathbf{u} - \tilde{\mathbf{U}}_l\right)(1 - 2\sigma)\right|. \tag{13}$$

In this case, the piecewise linear interpolant of the continuous function $\mathbf{u}$ is $l\mathbf{u}$, using the grid points of $\bar{\Omega}_l^N$. It needs to be observed that, in Sect. 3 (Step 2 of the procedure), the function $\mathbf{u}$ is continuous instead of discrete. The operator l and Lemma 4 are utilized to constrain the second phrase from (13) on the right. We obtain

$$\left| l\left(\mathbf{u} - \tilde{\mathbf{U}}_l\right)(1 - 2\sigma)\right| \le CN^{-1}\ln^2 N. \tag{14}$$

We apply the solution decomposition $\mathbf{u} = \mathbf{v} + \mathbf{w}$ for the first term in order to obtain

$$\left|(\mathbf{u} - l\mathbf{u})(1 - 2\sigma)\right| \le \left|(\mathbf{v} - l\mathbf{v})(1 - 2\sigma)\right| + \left|(\mathbf{w} - l\mathbf{w})(1 - 2\sigma)\right|. \tag{15}$$

Suppose $1 - 2\sigma$ lies in $\Upsilon = [t_i, t_{i+1}]$. Note that $\Upsilon \subset \bar{\Omega}_l$. The typical argument of piecewise linear interpolant lg, for every $g \in C^2(\Upsilon)$, produces

$$\left|(g - lg)(1 - 2\sigma)\right| \le Ch_i^2\|g^{(2)}\|_\Upsilon \quad \text{and} \quad \left|(g - lg)(1 - 2\sigma)\right| \le C\|g\|_\Upsilon. \tag{16}$$

Lemma 1 and the first bound of (16), $h_r \le CN^{-1}$ are used to obtain the value of the first phrase on the right side of (15)

$$\left|(\mathbf{v} - l\mathbf{v})(1 - 2\sigma)\right| \le Ch_i^2\|v^{(2)}\|_\Upsilon$$
$$\le CN^{-1}.$$

The second term of (15) has the following bounds:

$$\left|(\mathbf{w} - l\mathbf{w})(1 - 2\sigma)\right| \le C\|\mathbf{w}\|_\chi$$
$$\le Ce^{-\alpha\sigma/\epsilon}$$
$$\le CN^{-1},$$

where already applied the fact that $\sigma = \dfrac{2\epsilon}{\alpha}\ln N$. Hence, $\eta_{1-2\sigma} \le CN^{-1}\ln N$. Collecting the bounds for $\eta_{1-\sigma}$ and $\eta_{1-2\sigma}$, we obtain the intended result

$$\psi \le CN^{-1}\ln^2 N.$$

□

Theorem 6 *Suppose that the solution to (1) is $\mathbf{u}$ and that the kth iteration of the overlapping iterative method in Sect. 3 is $\mathbf{U}^{[k]}$. Then*

$$\|\mathbf{u} - \mathbf{U}^{[k]}\|_{\bar{\Omega}^N} \le C2^{-k} + CN^{-1}\ln^2 N.$$

Proof

$$\text{Defining,} \quad \tilde{U}(t_i) = \begin{cases} \tilde{U}_r(t_i), & \text{if } t_i \in \bar{\Omega}_r^N \setminus \bar{\Omega}_l \\ \tilde{U}_l(t_i), & \text{if } t_i \in \bar{\Omega}_l^N. \end{cases}$$

An inequality triangle is used to obtain

$$\|\mathbf{u} - \mathbf{U}^{[k]}\|_{\bar{\Omega}^N} \leq \|\mathbf{u} - \widetilde{\mathbf{U}}\|_{\bar{\Omega}^N} + \|\widetilde{\mathbf{U}} - \mathbf{U}^{[k]}\|_{\bar{\Omega}^N}. \tag{17}$$

Using Lemma 4, in (17), the right-hand first expression has reached its limit. We give the note that follows before pushing for a bound on the second term:

$$\omega^{[k]} = \|\widetilde{\mathbf{U}} - \mathbf{U}^{[k]}\|_{\bar{\Omega}^N},$$
$$\eta^{[k]} = \|(\widetilde{\mathbf{U}}_r - l\mathbf{U}_l^{[k-1]})(1 - 2\sigma)\|_\infty.$$

Given that $\mathbf{U}^{[0]}(t_i) = \mathbf{0}$, $0 < t_i < 1$, we can demonstrate that $\eta^{[1]} \leq C$ by using Lemma 1 and the discrete maximum principle. We have

$$L^N\left(\widetilde{\mathbf{U}}_r - \mathbf{U}_r^{[1]}\right) = \mathbf{0} \text{ in } \Omega_r^N, \quad \left|\left(\widetilde{\mathbf{U}}_r - \mathbf{U}_r^{[1]}\right)(1 - 2\sigma)\right| = \left|\left(\widetilde{\mathbf{U}}_r - l\mathbf{U}_l^{[0]}\right)(1 - 2\sigma)\right| \leq \eta^{[1]},$$
$$\mathbf{B}_1\left(\widetilde{\mathbf{U}}_r - \mathbf{U}_r^{[1]}\right)(1) = \mathbf{0}.$$

Applying discrete maximum principle for the operator L^N on $\bar{\Omega}_r^N$ and using the suitable barrier function, it can be deduced that

$$\|\widetilde{\mathbf{U}}_r - \mathbf{U}_r^{[1]}\|_{\bar{\Omega}_r^N} \leq \frac{\mathbf{C}}{2}\eta^{[1]}. \tag{18}$$

Consequently

$$\|\widetilde{\mathbf{U}}_r - \mathbf{U}_r^{[1]}\|_{\bar{\Omega}^N \setminus \bar{\Omega}_m} \leq \frac{\mathbf{C}}{2}\eta^{[1]}. \tag{19}$$

Also,

$$L^N\left(\widetilde{\mathbf{U}}_l - \mathbf{U}_l^{[1]}\right) = \mathbf{0} \text{ in } \Omega_l^N, \quad \mathbf{B}_0\left(\widetilde{\mathbf{U}}_l - \mathbf{U}_l^{[1]}\right)(0) = \mathbf{0},$$

$$\left|\left(\widetilde{\mathbf{U}}_l - \mathbf{U}_l^{[1]}\right)(1 - \sigma)\right| = \left|(\widetilde{\mathbf{U}}_l - l\mathbf{U}_r^{[1]})(1 - \sigma)\right|$$
$$\leq \eta_{1-\sigma} + \frac{\mathbf{C}}{2}\eta^{[1]}.$$

In this instance, it has been used that the mesh point of $\bar{\Omega}_r^N$ is $1 - \sigma$.

Therefore, for the operator L^N, we obtain, by the discrete maximum principle,

$$\|\widetilde{\mathbf{U}}_l - \mathbf{U}_l^{[1]}\|_{\bar{\Omega}_l^N} \leq \eta_{1-\sigma} + \frac{\mathbf{C}}{2}\eta^{[1]}. \tag{20}$$

We now estimate $\eta^{[2]}$, which will be used to set the constraint for $\omega^{[2]}$, the second iteration. Every term in $\eta^{[2]}$ is evaluated independently for this purpose. When a triangle inequality is applied,

$$
\begin{aligned}
\left|(\tilde{\mathbf{U}}_r - l\mathbf{U}_l^{[1]})(1 - 2\sigma)\right| &\leq \left|(\tilde{\mathbf{U}}_r - l\tilde{\mathbf{U}}_l)(1 - 2\sigma)\right| + \left|(l\tilde{\mathbf{U}}_l - l\mathbf{U}_l^{[1]})(1 - 2\sigma)\right| \\
&= \left|(\tilde{\mathbf{U}}_r - l\tilde{\mathbf{U}}_l)(1 - 2\sigma)\right| + \left|(\tilde{\mathbf{U}}_l - \mathbf{U}_l^{[1]})(1 - 2\sigma)\right| \\
&\leq \eta_{1-2\sigma} + \frac{C}{2}\eta^{[1]} + \eta_{1-\sigma},
\end{aligned}
$$

where the operator l and (20) stability have been utilized.

After collecting the estimates mentioned above, we get

$$
\eta^{[2]} \leq \frac{C}{2}\eta^{[1]} + \eta_{1-\sigma} + \eta_{1-2\sigma}.
$$

Thus, in the initial iteration, we get

$$
\max\left\{\eta^{[2]}, \omega^{[1]}\right\} \leq \frac{C}{2}\eta^{[1]} + \psi, \quad \text{where } \psi = \eta_{1-\sigma} + \eta_{1-2\sigma}.
$$

Now, if we continue to iterate the aforementioned argument, we get

$$
\max\left\{\eta^{[k+1]}, \omega^{[k]}\right\} \leq \frac{1}{2}\eta^{[k]} + \psi
$$

Considering the inequalities previously mentioned, we obtain $\eta^{[k]} \leq 2^{-(k-1)}\eta^{[1]} + 2\psi$, and hence

$$
\omega^{[k]} \leq 2^{-k}\eta^{[1]} + 2\psi.
$$

In the end, we get the bound on $\omega^{[k]}$ using Lemma 5, which is obtained from the inequality above,

$$
\|\tilde{\mathbf{U}} - U^{[k]}\|_{\bar{\Omega}^N} \leq C2^{-k} + CN^{-1}\ln^2 N. \tag{21}
$$

Thus, we obtain the necessary result from (17) by applying Lemma 4 and (21). $\quad\square$

Lemma 7 *Let $\mathbf{U}^{[k]}$ represent the discrete method's kth iteration, described in Sect. 3. Then there is a C such that*

$$
\|\mathbf{U}^{[k+1]} - \mathbf{U}^{[k]}\|_{\bar{\Omega}^N} \leq Cv^k \quad \text{where} \quad v = \left(1 + \frac{\sigma\alpha}{2\epsilon N}\right)^{-N/2} < 1,
$$

and C is independent of k and N. Additionally, if $\sigma = \dfrac{2\epsilon}{\alpha}\ln N$ then $v \leq 2N^{-1/2}$.

Proof The present theorem can be proved by using the proof strategy described in [16]. $\quad\square$

The following theorem, which combines Lemma 7 and Theorem 6, contains the major result of this article: Results indicate that obtaining almost first-order convergence requires only two iterations.

Theorem 8 *Let* $\mathbf{U}^{[k]}$ *be the kth iteration of the discrete method given in Sect. 3, and let* $\mathbf{u}$ *be the solution to (1). If* $\sigma = \dfrac{2\epsilon}{\alpha} \ln N$ *and* $N > 2$, *then*

$$||\mathbf{U}^{[k]} - \mathbf{u}||_{\bar{\Omega}^N} \leq CN^{-k/2} + CN^{-1} \ln^2 N$$

and C is not dependent on k, ϵ, *and* N.

Proof The proof approach explained in [16] can be used to prove that theorem. □

5   Numerical Results

To illustrate the theoretical outcomes of the SPBVPs, we examine one case in this section (1). It is assumed that the iterative procedure's stopping condition is $||\mathbf{U}^{[k+1]} - \mathbf{U}^{[k]}||_{\bar{\Omega}^N} \leq 10^{-14}$. Given the exact solution u_j on the mesh $\bar{\Omega}^N$ let U_j^N be a numerical estimation, and N represent the total number of grid points. When writing U_j^N on the last iteration, we typically omit the superscript k. Given an limited number of values, $\epsilon = \{10^{-1}, 10^{-3}, 10^{-5}, \cdots, 10^{-19}\}$. For $j = 1, 2$, we determine the greatest point-wise errors in the two mesh differences.

$$||U_j^N - u_j||_{\Omega^N} \approx D_{\epsilon,j}^N := ||U_j^N - \bar{U}_j^{2N}||_{\Omega^N}, \quad D_j^N = \max_{\epsilon} D_{\epsilon,j}^N,$$

where $\bar{U}_j^{2N}$ is the computational answer on a grid with the same transitional points, but 2N intervals in each subdomain. The rate of convergence is determined using these values.

$$p_j^N = \log_2\left\{\frac{D_j^N}{D_j^{2N}}\right\}, \quad j = 1, 2.$$

For a set of values of N and ϵ, Table 1 tabulates the calculated maximum point-wise errors D_j^N ($j = 1, 2$), the evaluated order of convergence p_j^N ($j = 1, 2$), and k (the number of computed iterations). Nodal error plots are displayed (Fig. 2). It is evident that the errors diminish with increasing N and are not depending upon the singular perturbation parameter ϵ. These approaches are generally used in conjunction with a ln N factor, and the computed convergence rates are almost first order.

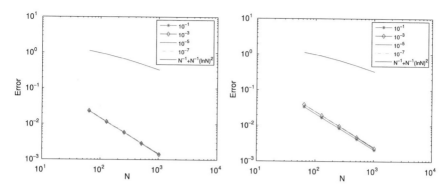

Fig. 2 Nodal error in U_1 and U_2 components of Example 1

Example 1 Consider the SSPCDP

$$-\epsilon u_1''(t) + (1+t)u_1'(t) + 3u_1(t) - u_2(t) = 2;$$
$$-\epsilon u_2''(t) + (1 + \frac{t^2}{2})u_2'(t) - u_1(t) + 2u_2(t) = \frac{e^t}{3},$$
$$u_1(0) - u_1'(0) = 1, \quad 3u_1(1) + \epsilon u_1'(1) = 1,$$
$$u_2(0) - u_2'(0) = 2, \quad 3u_2(1) + \epsilon u_2'(1) = 2.$$

Table 1 presents the numerical results.

Table 1 For Example 1, the values of the solution components U_1 and U_2 are D_1^N, p_1^N and D_2^N, p_2^N, respectively

Mesh point value (N)					
ϵ	64	128	256	512	1024
D_1^N	2.2510e−003	1.1676e−003	5.9458e−004	3.0001e−004	1.5069e−004
p_1^N	**9.4702e−001**	**9.7360e−001**	**9.8686e−001**	**9.9343e−001**	–
k	2	2	2	2	2
D_2^N	1.9548e−003	9.9620e−004	5.0286e−004	2.5263e−004	1.2662e−004
p_2^N	**9.7251e−001**	**9.8628e−001**	**9.9313e−001**	**9.9652e−001**	–
k	2	2	2	2	2

6 Conclusion

We have developed first-order convergence of overlapping iterative numerical technique for SSPCDEs. It seems that a planned discrete iterative approach yielded numerical approximations that converged to the exact answer in the maximum norm. It is demonstrated that this convergence is first order. It should be noted that the $N^{-k/2} + N^{-1} \ln^2 N$ term dominated the error bound for $k \geq 2$, according to Theorem 8. To achieve the required accuracy, two iterations have become sufficient. The theoretical conclusions have been tested using numerical experimentation.

References

1. Christy Roja J, Tamilselvan A, Geetha N (2020) An analysis of overlapping Schwarz method for a weakly coupled system of singularly perturbed convection-diffusion equations. Int J Numer Methods Fluids 92(6):528–544
2. Kumar S, Kumar M (2015) An analysis of overlapping domain decomposition methods for singularly perturbed reaction-diffusion problems. J Comput Appl Math 281:250–262
3. Kokotović PV (1984) Applications of singular perturbation techniques to control problems. SIAM Rev 26(4):501–550
4. Turkyilmazoglu M (2010) Series solution of nonlinear two-point singularly perturbed boundary layer problems. Comput Math Appl 60(7):2109–2114
5. Turkyilmazoglu M (2011) Analytic approximate solutions of parameterized unperturbed and singularly perturbed boundary value problems. Appl Math Modell 35(8):3879–3886
6. Basha PM, Shanthi V (2015) A numerical method for singularly perturbed second order coupled system of convection-diffusion robin type boundary value problems with discontinuous source term. Int J Appl Comput Math 1:381–397
7. Rao SCS, Chawla S (2020) Robin boundary value problems for a singularly perturbed weakly coupled system of convection-diffusion equations having discontinuous source term. J Anal 28:305–321
8. O'Riordan E, Stynes J, Stynes M (2008) A parameter-uniform finite difference method for a coupled system of convection-diffusion two-point boundary value problems. Numer Math Theor Meth Appl 1(2):176–197
9. Liu L-B, Liang Y, Bao X, Fang H (2021) An efficient adaptive grid method for a system of singularly perturbed convection-diffusion problems with robin boundary conditions. Adv Differ Equ 2021(1):1–13
10. Priyadharshini RM, Ramanujam N (2013) Uniformly-convergent numerical methods for a system of coupled singularly perturbed convection-diffusion equations with mixed type boundary conditions. Math Modell Anal 18(5):577–598
11. Tamilselvan A, Ramanujam N, Priyadharshini RM, Valanarasu T (2010) Parameter-uniform numerical method for a system of coupled singularly perturbed convection-diffusion equations with mixed type boundary conditions. J Appl Math Inform 28(1):109–130
12. Farrell P, Hegarty A, Miller JM, O'Riordan E, Shishkin GI (2000) Robust computational techniques for boundary layers. CRC Press
13. Miller JJ, O'Riordan E, Shishkin GI (1996) Fitted numerical methods for singular perturbation problems: error estimates in the maximum norm for linear problems in one and two dimensions. World scientific
14. Priyadharshini RM, Ramanujam N, Tamilselvan A (2009) Hybrid difference schemes for a system of singularly perturbed convection-diffusion equations. J Appl Math Inform 27(5):1001–1015

15. Mythili Priyadharshini R, Ramanujam N, Shanthi V (2009) Approximation of derivative in a system of singularly perturbed convection-diffusion equations. J Appl Math Comput 29(1–2):137–151

16. Roja JC, Tamilselvan A (2018) Schwarz method for singularly perturbed second order convection-diffusion equations. J Appl Math Inform 36(3):181–203

17. MacMullen H, Miller J, O'Riordan E, Shishkin G (2001) A second-order parameter-uniform overlapping Schwarz method for reaction-diffusion problems with boundary layers. J Comput Appl Math 130(1–2):231–244

18. MacMullen H, O'Riordan E, Shishkin G (2002) The convergence of classical Schwarz methods applied to convection-diffusion problems with regular boundary layers. Appl Numer Math 43(3):297–313

Correction to: Design of Microstrip Rectangular Dual Band Antenna for MIMO 5G Applications

M. Jayasudha, P. Ranjitha, Senthil Kumaran R, V. Jayasudhan, and S. Yuvaraj

Correction to:
Chapter 7 in: D. Giri et al. (eds.), *Proceedings of the Tenth International Conference on Mathematics and Computing*, Lecture Notes in Networks and Systems 964, https://doi.org/10.1007/978-981-97-2066-8_7

In the original version of the book, the following belated corrections have been incorporated: The author name "R. Senthil Kumaran" has been changed to "Senthil Kumaran R" in the Frontmatter and in Chapter 7. The book has been updated with the changes.

The updated version of this chapter can be found at
https://doi.org/10.1007/978-981-97-2066-8_7

Author Index

Printed in the United States
by Baker & Taylor Publisher Services